'A master class in understanding the use of production technology as an arena of musical practice.

In this fascinating and highly accessible account, Thomas Brett proves his skills not just as a music maker, but also as a critical thinker on the subject of music production as a cultural space. He experiments and samples widely. He blends. He makes forays into different disciplines his own, creating an ongoing series of metaphors that repeatedly deliver insights into music production as a field. More than simply offering a masterful introduction to production technology, Brett teaches us to consider its sociological context and cultural resonances. His notion of the producer-fan helps us rethink the out-dated dichotomy that artificially separates fans from music makers, when we know those roles can and do overlap.

You could not ask for a better guide on your quest to fully understand the craft.'

Mark Duffett, *University of Chester*

The Creative Electronic Music Producer

The Creative Electronic Music Producer examines the creative processes of electronic music production, from idea discovery and perception to the power of improvising, editing, effects processing, and sound design.

Featuring case studies from across the globe on musical systems and workflows used in the production process, this book highlights how to pursue creative breakthroughs through exploration, trial and error tinkering, recombination, and transformation.

The Creative Electronic Music Producer maps production's enchanting pathways in a way that will fascinate and inspire students of electronic music production, professionals already working in the industry, and hobbyists.

Thomas Brett is an electronic music producer, with a PhD in ethnomusicology from New York University. His essays and book reviews have appeared in the journals *Popular Music* and *Popular Music and Society*, as well as edited collections by Routledge, and Oxford and Cambridge University presses. He has released numerous electronic music recordings, and his compositions for marimba and vibraphone have been performed internationally by soloists and university ensembles. Thomas has played percussion on Broadway since 1997 and writes about music at brettworks.com.

Perspectives on Music Production

Series Editors:
Russ Hepworth-Sawyer
York St John University, UK
Jay Hodgson
Western University, Ontario, Canada
Mark Marrington
York St John University, UK

This series collects detailed and experientially informed considerations of record production from a multitude of perspectives, by authors working in a wide array of academic, creative and professional contexts. We solicit the perspectives of scholars of every disciplinary stripe, alongside recordists and recording musicians themselves, to provide a fully comprehensive analytic point-of-view on each component stage of music production. Each volume in the series thus focuses directly on a distinct stage of music production, from pre-production through recording (audio engineering), mixing, mastering, to marketing and promotions.

Pop Music Production
Manufactured Pop and Boy Bands of the 1990s
Phil Harding
Edited by Mike Collins

Cloud-Based Music Production
Sampling, Synthesis, and Hip-Hop
Matthew T. Shelvock

Gender in Music Production
Edited by Russ Hepworth-Sawyer, Jay Hodgson, Liesl King and Mark Marrington

Mastering in Music
Edited by Russ Hepworth-Sawyer and Jay Hodgson

Innovation in Music
Future Opportunities
Edited by Russ Hepworth-Sawyer, Justin Paterson and Rob Toulson

Recording the Classical Guitar
Mark Marrington

The Creative Electronic Music Producer
Thomas Brett

3-D Audio
Edited by Justin Paterson and Hyunkook Lee

For more information about this series, please visit: www.routledge.com/Perspectives-on-Music-Production/book-series/POMP

The Creative Electronic Music Producer

Thomas Brett

Routledge
Taylor & Francis Group

LONDON AND NEW YORK

First published 2021
by Routledge
2 Park Square, Milton Park, Abingdon, Oxon OX14 4RN

and by Routledge
605 Third Avenue, New York, NY 10158

Routledge is an imprint of the Taylor & Francis Group, an informa business

British Library Cataloguing-in-Publication Data
A catalogue record for this book is available from the British Library

Library of Congress Cataloging-in-Publication Data
A catalog record for this book has been requested

ISBN: 978-0-367-90080-9 (hbk)
ISBN: 978-0-367-90079-3 (pbk)
ISBN: 978-1-003-02246-6 (ebk)

Typeset in Times New Roman
by Apex CoVantage, LLC

Contents

About the author

Thomas Brett is a musician and writer who holds a PhD in ethnomusicology from New York University. His essays and book reviews have appeared in the journals *Popular Music* and *Popular Music and Society* as well as edited collections by Routledge, Oxford, and Cambridge University presses. He has released numerous electronic music recordings, and his compositions for marimba and vibraphone have been performed internationally by soloists and university ensembles. Thomas has played percussion on Broadway since 1997 and writes about music at brettworks.com.

Figures

Acknowledgments

Thank you to the editorial staff at Routledge for their guidance; the *Perspectives On Music Production* series editors Russ Hepworth-Sawyer, Mark Marrington, and Jay Hodgson for their ideas, feedback, and support for this project; Russell Hartenberger for reading and commenting on a manuscript draft; Ryan Lee West for graciously letting me use one of his track sketch photos; and Jack Mansager for his diagram rendering. This book is dedicated to the producers, composers, musicians, and software discussed in this book whose music, ideas, and potentials have been sources of inspiration: Ableton, Actress, Afrika Bambaataa, Akira Rabelais, Alva Noto, Amon Tobin, Andrew Huang, Aphex Twin, Arovane, Arturia, Autechre, Arca, Arvo Pärt, Autechre, Au5, Beatrice Dillon, Ben Lukas Boysen, Biosphere, Boards Of Canada, Boris Brejcha, Boxcutter, Brian Eno, Burial, Caterina Barbieri, Clark, Culprate, deadmau5, Deru, DJ Kool Herc, Dyalla, Floating Points, Flying Lotus, Four Tet, Frequent, Giorgio Moroder, Glenn Gould, G Jones, Goodhertz, Guy Sigsworth, Hank Shocklee, Harold Budd, Harold Melvin & The Blue Notes, 4Hero, Holly Herndon, Ikonika, Imogen Heap, Isao Tomita, Ital Tek, Jamie Lidell, J Dilla, Jean-Michel Jarre, Jimmy Bralower, Jlin, John Cage, John Frusciante, Jon Hopkins, Joris Voorn, JVNA, Kaitlyn Aurelia-Smith, Kara-Lis Coverdale, Keith Jarrett, King Tubby, Lanark Artefax, Laurel Halo, Laxity, Logos, Lorenzo Senni, Loscil, Mark Fell, Matmos, Max Loderbauer, Miles Davis, Modartt, Mr. Bill, New Order, Nils Frahm, Noisia, Nosaj Thing, Objekt, Ólafur Arnalds, Oneohtrix Point Never, Philth, Phuture, Pierre Schaeffer, Pole, Prince Paul, Public Enemy, Ransby, r beny, Ricardo Villalobos, Rival Consoles, Rob Clouth, Robert Henke, Roni Size & Reprazent, Sam Barker, Skee Mask, Skrillex, SOPHIE, Squarepusher, Steinberg, Stenny, Steve Duda, Steve Hauschildt, Steve Reich, Ted Macero, Terry Riley, The Winstons, Timbaland, Tim Hecker, Tipper, TM404, Tom Cosm, Tom Holkenborg, Trevor Horn, U-he, Virtual Riot, and William Basinski.

Introduction
Electronic music production as infinite game

Picture an electronic music producer at work. S/he sits with headphones on, facing the glow of a computer screen, a hand on the trackpad, pointing and clicking, drawing in notes, dragging parts around, making adjustments inside music software, listening to what happens. Working with synthesis and samples, beats and rhythms, textures, soundscapes, and ambiances, s/he's building a track of shape-shifting sonics whose virtual world of timbres and relations goes beyond the acoustic. The producer is in Accra or Berlin, Chile or Toronto, designing sounds by collaborating with his/her tools, shaping the music, and evolving perception. The track could be pop or hip hop, floating ambient or kinetic bass, but making it—*producing* it—is an experimental, future-oriented process that takes risks, that runs counter to convention (Demers, 2010: 7). This is the producer's Quest: to create sounds, structures, and feelings that have never been configured this way before.

This book is about the craft of electronic music production and how producers think about it. Its impetus is my own curiosity about how other producers approach production. The chapters herein offer reflections on practitioners' experiences, configuring a mix of history, how-to guide, auto-ethnography, and idea sampling from a broad span of practices about the dynamics of invention and artistry in electronic music production. The book explores the uncertainties and discoveries of the production process, particularly as it relates to using DAW (digital audio workstation) software, the most significant development in music production since the invention of tape recording in the 1930s. DAWs such as Live, Logic, Cubase, FL Studio, Pro Tools, Reason, BitWig, Reaper, and others turn computers into studios, open up production to an ever-growing community of practitioners, and give musicians a dynamic meta-tool with which to compose. It is within this context that I explore electronic music production by considering the interactions between producers' thinking, their tools, their workflows for using these tools, and the music that results. From sound design to improvising, from editing to arranging and mixing, the craft of production lives in the production process itself.

Producing electronic music is open-ended—a kind of infinite game that is seemingly without end. The reason for this is that, at any point in the process of making a track, there is so much a producer can do. Whether recording audio, synthesizing sounds, sequencing a groove, or chopping samples, the producer has options upon options as to *what's next*. This open-endedness sometimes creates a sense that one's music will never have definitive point of being finished. In the DAW, a track is an elastic composition in flux, always one mouse click away from a variation or a reworking, as fungible as our understanding of its potential to become something else. It is for this reason that the perceptual aspect of production is key: the shape our music takes is inextricably connected to what we notice and how we respond to that noticing. The craft of electronic music production, then, is an experience that expands in proportion to our interaction with the music in progress: the music takes form as we think through what to do (or not to do) next, react to happy accidents, and deliberately refine its sounds. When you are producing, the uncertainties you encounter are your navigational system for figuring out where to go next. *What's next?* is the most urgent question, a springboard for forward motion as you leap from how the music sounds now to the future of what it might become.

A perception and decision-making Quest for tacit knowledge

Philosophically, this book is guided by three ideas. The first is that how you interact with, and respond to, your musical system (a set of tools) shapes your music. How you produce is influenced by your prior musical experience, your go-to sounds (whether you use presets, loops, or make your own sounds), and the musical controllers such as keyboards or drum pads you use to get ideas into the computer. But experience and equipment merely support the most incisive tools you already possess: your perception and decision-making. Your perceptions shape how you understand and interpret what you hear, and your decision-making builds on this awareness to bring you to the next steps.

The book's second contention is that music production is a discovery adventure or, more grandly, a Quest, a word from the Latin verb *quaerere* which means "to ask, seek." Your Quest is a seeking to discover ways of producing that make sense of your experience and distill your disparate influences and obsessions into your own sound. Producing music is a gradual process, which means that a production workflow that creates meaningful sonic results that are authentically yours is earned through experience. For producer Prince Paul (Paul Edward Huston), production is asking questions:

> I think that's the best thing in production when you ask yourself questions, "Can I make this go backwards but only the snares go forward?" And then you start to find out the answers to those questions and that, I think, enables greater production.
>
> (Schmidt, 2003)

The book's third contention, which may sound obvious to some readers, is that the craft of electronic music production is learned through practice. The producer learns from doing, and more specifically, from iterations over time. It is from doing and iterating over time that I have learned about a software's functionalities and how to make sounds feel one way or another. What may not be obvious is that it is only from practice that a producer acquires *tacit knowledge*. Tacit means understood or implied without being stated, and tacit knowledge is knowledge we arrive at from personal experience that is difficult to verbalize and explain. The concept was developed by Michael Polyani, who summarizes it as "we can know more than we can tell" (Polyani, 1966: 4). This book tries to tell more than I know by gathering and amplifying the tacit knowledge that guides the practices of electronic music producers.

Following Prince Paul, each time you produce you set up a feedback loop between the questions you ask about sound and the answers you receive in sound, in effect developing your tacit knowledge. This book is neither a *how-to* book nor does it proscribe a one-approach-fits-all way to electronic music production, because such an approach does not exist. Moreover, even though there are many established methods for how to make already recognized sounds (e.g. *lo-fi* drums, or *supersaw* leads), I urge you to think of your production Quest as a way to evolve your own processes and workflows for creating sounds *you* would like to hear—or didn't know you wanted to hear until you heard them, just now. In this sense, *The Creative Electronic Music Producer* takes a *How about?* approach. As we will learn in the chapters that follow, there are so many ways a producer can move. Some producers build upon beats, some work with melodies or drones, some create sample libraries that become music, some sound design as they go along, some prefer hardware synthesizers, some sample from field recordings, and some do a bit of everything. But since no production approach is comprehensive and not all approaches may speak to you, there is no reason to commit to a single way of working. A distinguishing quality of attuned producers is that they are tacit knowledge-based "creative technicians" who build tracks with whatever they have at hand, paying close attention and making careful decisions based upon what they hear (Harding, 2020: 58). *Paying close attention* is key because it leads us to compelling sounds and helps us decide where to take them next. In sum, electronic music production is a call and response between you, the producer, and a set of tools you have configured as a musical system. Production lives in the space between the music and our responding to its sounds. It lives in the craft of uncovering in our tacit knowledge what we didn't know we already know.

From outer inspiration to inner spur

I am neither a newcomer to electronic music production nor an especially advanced practitioner. In 2020, I made a recording, *Plentitudes*, for marimba and electronic sounds. I had made other recordings before this one, but this project was more involved, combining my marimba playing with beats, melodies, bass lines, and processing. I took notes on my workflow and insights, my mistakes and frustrations, and these notes were a starting point for the topics explored in this book. As the music took shape, I wanted to connect my project with the practices of other electronic music producers by trying to understand the structure of their experience. I was interested in learning about how other producers arrive at what they know, how they apply that knowledge in their production work, and the dynamic relationship between their knowing and their sounds.

Like many producers, I am continually listening to the music of others and wondering how and why they made it sound the way it does. Interesting and enchanting sounds motivate me, and I have my antennae out for production concepts, tips, and hacks I might learn from. I read interviews, watch production tutorials on YouTube (more about which in this book's Interlude), browse gear reviews and discussion threads, and sometimes even read software manuals. This information about music production culture as a field of practice offers a sense of how others work and what is possible to do with production tools. We can call this *outer inspiration*, a tuning into a broader music electronic production community's co-evolution with innovations in technology, fashions of musical style, and producers' responding to these shifts in their practices. But understanding how others work is only part of acquiring tacit knowledge about production.[1] The other part is discovering and refining one's own production workflows. We can call this *inner spur*. Compressed to its essentials, the inner spur of production has three components: make music, learn from your making (especially from your mistakes), and iterate (repeat) the process. Think of your producing as a complex system that creates feedback loops between the sounds you make and how you respond to them. In this book, then, I urge producers to consider the possibility that more than specific concepts, tips, or hacks, *the feedback loops you create yourself are the engines of your producing*. Feedback loops are learning machines we devise that spur us towards insights. Which brings us to the topic of creativity.

Creativity, the experimentalist's mindset, and making many small mistakes

Creativity—"the generation of high-quality, original, and elegant solutions to complex, novel, ill-defined problems" (Mumford et al., 2012)—is a fundamental tone running through this book. The word is often used to describe using imagination or original ideas to produce something *beautiful* or *innovative* or *of value* like a piece of music, a dance, a theorem, a novel, a sculpture (cf. Boden, 2004: 1). As listeners and fans of electronic music production, we find a track creative when we notice its sounds doing things that haven't been done quite like this before. We hear new timbral, rhythmic, or melodic worlds; a spin on a musical style that is some way unique; or we hear a producer approaching familiar sounds from new angles to create an unusual beauty or unexpected feeling. We might say that producers think about creativity insofar as they seek the original and unconventional over what has already been done, continually refining their craft in pursuit of new sounds.

It is commonplace to understand creativity in terms of expression, because our made things have an expressive form or intent. But, as I hope to show, we might also understand creativity as a way of responding to a situation that has already been set in motion. In fact, for many producers, *creativity is an approach to practice* that engages with complexity and welcomes surprise moments. In electronic music production, *creativity happens in the interaction between the producer and the musical system in which s/he works*. The components of a musical system differ for each producer, but ideally, *the system has the potential for some degree of complexity and nonlinearity*, meaning that it can generate outputs different from, and greater than, its inputs. Like a microphone placed in front of an amplifier that generates a screeching feedback tone, a complex system connects a producer's workflow and its resulting sounds in feedback loops that can cascade in nonlinear ways over time. Just as the tone is picked up by the microphone and amplified back into it, feedback loops carry a producer around and around the circuit of a system. By trying things out, making small changes, swapping sounds, routing in effects, and so on, any one of a producer's creative decisions can have huge output implications. As complexity theorist Donella H. Meadows notes, "complex and delightful patterns can evolve from quite simple sets of rules" (Meadows, 2008: 159). From a complexity perspective then, producing electronic music is a way of responding to nonlinear, turbulent, and dynamic processes never entirely under the producer's control. As we will learn in Chapter 2, the DAW or any other configuration of technologies used as a musical system is a powerful tool for production discoveries through complexity.

This book reframes conventional thinking about creativity in electronic music production and its implications for how we work, drawing on insights from producers as well as the work of practitioners outside of the musical realm.[2] Perhaps its most vital production insight is that *creativity does not emerge from random anywheres, but from a system in which it can flourish*. Producers arrive at such systems through experimentation, random discoveries, and iteration. As we adopt the mindset that creativity in production is not inspiration per se, but experimentation and discovery, not waiting for good ideas to appear but instead playing with whatever one has (a sound, a pattern, a sample, a parameter, a malfunction), we position ourselves to discover novel ways of combining and transforming our materials to make new sounds. Adopting an experimentalist's tinkering and iterative mindset, we notice creative moments emerging almost as if by their own powers—as a by-product of our having spent time trying things out. Whether you are new to electronic music production or an experienced practitioner, give yourself permission to continually experiment, repeatedly make many small mistakes, and disrupt your habitual ways of working.[3]

Three aims

In the chapters that follow I interweave my work on *Plentitudes* with the history of electronic music production and the practices of other producers whose sounds, approaches, ideas, and sense of adventure have inspired me. (Examples of their music are provided in this book's Suggested Listening.) The chapters are roughly chronological, each one considering a different facet of producing electronic music: setting up a musical system, choosing sounds, improvising and recording ideas, rhythm programming, disruption, editing, arranging, and mixing. While these production tasks are not necessarily distinct—in practice they often overlap and can happen in any order—by treating them separately I hope to convey the essentials of their workflows, guiding concepts, and links to broader themes in the history of music production. In sum, I have three aims:

1   to suggest ways of thinking holistically about creativity in electronic music production as a complex and dynamic system by describing the open-ended, ever-changing encounter between the music producer and sound's enchanting landscape of possibilities
2   to distill lessons from experiences of adventure, uncertainty, serendipity, complexity, and feedback loops that arise when producing electronic music
3   to suggest ways of moving production forward through iterative, trial and error-based tinkering, improvising, exploring, recombining, and transforming

Think of this book as a mixing console through which to combine ideas about electronic music production. Think of its chapters as channel faders and effects send knobs on a mixing console, with each fader on the console raising the volume on a different theme of the text, each knob routing these themes to the thoughts of other producers and back again, and blending this shared tacit knowledge about production into a composite mix. Like a reverb+delay+distortion chain of effects processing a sound, *The Creative Electronic Music Producer* routes sounds into words, listens for ghost notes and reverb tails, and explains production processes as circuits of understanding.

Chapter outline

This book's chapters combine a mix of reflections, examples, and case studies to consider the practice of electronic music production through a series of related topics. Chapter 1 works backwards from our current production moment to trace a brief history of audio production in which I explore the figure of the music producer, the studio as compositional tool concept, and the advent of MIDI, digital sampling, and the DAW. A case study on Ableton Live, the most influential DAW of the past two decades, explains how the software encourages an experimentalist mindset in electronic music production. The chapter concludes with a brief discussion of basic DAW topology and concepts to orient the novice producer.

Chapter 2 explains the importance of configuring complex musical systems and workflows for interacting with them. I provide numerous examples of such systems, with a case study on Brian Eno's early systems of sonic treatments and generative loop machines. The chapter concludes with suggestions for setting up a music production system, controllers, templates and sound sets, and how to begin the process.

Chapter 3 considers the beginnings of a production project. I connect my experience working on *Plentitudes* with insights from other composers and producers, describing broad concepts to consider when starting a project that help the producer interact with a musical system. The concepts are: devising the appropriate system for a musical gesture, discovering the dialectics among our ideas and our musical system and how they "talk" to one another, focusing on a small thing to discover everything musical in it, production placeholders and through-lines, finding permutations and combinations of a limited sound set to create series of variations, and letting tools and their workflows guide and outsource creativity. I also consider how improvisation shapes many facets of production: finding interesting sounds, happening upon accidental counterpoint, being lost and getting un-lost, using call and response, emulating the sound of a virtual band, and capturing maximum musicality. I conclude by discussing the importance of mindset, the value of seeking out the unusual to allow for interesting production discoveries, and steps for resisting the musically predicable.

Chapter 4 examines presets and sound design in electronic music production. I trace a history of preset sounds from the one-touch rhythms on home organs, 1980s synthesizer patches, to today's VST plug-ins, sample packs, and production "composing kits," with a case study on Spitfire Audio's *Composer Toolkits*. Building upon my experience designing sounds to accompany my marimba playing, I consider how producers negotiate limited sound sets, production disconnects, the 80% lesson, synthesis thinking, the acoustic sound ideal, and sonic uncanny valleys. In sum, despite the plethora of sounds available to producers today, designing one's own sounds is the surest path to finding one's own production voice.

Interlude: YouTube electronic music tutorials considers representations of electronic music production and fandom in online video tutorials. First I explain the role fandom plays in electronic music production, focusing on the specialized activity of *producer-fans* who are amateur producers themselves. Some of these producer-fans create production tutorials, while others watch the tutorials and leave comments. This fan activity offer us a way to assess how knowledge about music production is conveyed and consumed. Next I turn to five production tutorial examples from the YouTube channels of Tom Cosm, Andrew Huang, Bill Day (Mr. Bill), Joel Zimmerman (deadmau5), and Christian Valentin Brunn (Virtual Riot). These videos offer aspiring producers a plethora of ideas about techniques and workflows for sound design, such as Brunn's six production concepts: *Create the MIDI part first, Wet and dry contrast, Layering, Layers of rhythmic movement, Random-sounding parts*, and *Make boring sounds more interesting*. In sum, producers' workflows, techniques, commentary, and choice of sounds give us a sense of tacit production knowledge in action, inspiring us on our own production Quests.

Chapter 5 traces a history of rhythm programming, one of the most intricate and impactful components of electronic music production. The chapter explores how producers from the 1980s onwards used drum machines, pad-based Grid controllers, and then DAWs to program rhythms. I show how producers' encounters with these technologies shaped how they think about rhythm and beat construction. Next I discuss the anatomy of rhythmic grooves as constructions with a limited set of sounds in tension and opposition that develop over time, and offer five rhythm programming examples. I conclude with five general rhythmic principles applicable to any musical context: play grooves, layer rhythmic parts, go beyond drum fills and short-term phrasing, think in timelines and pulsation, and edit sounds and rhythms.

Chapter 6 examines moments of disruption in the music production process that can change the direction of a track-in-progress. I explore the implications of disruption in the form of linear and nonlinear ways of working, creating variations from a single sound, reincorporating mistakes, applying effects in layers, and musical transitions. Two case studies, on the musics of producers Sam Barker and Lorenzo Senni, consider the disruptive techniques of muting four-on-the-floor beats and isolating build-ups. I argue that disruption requires the producer to subvert a workflow, question methods, deconstruct or distort too-clean or otherwise predictable sounds, re-pitch and re-arrange sounds, or apply some other form of change.

Chapter 7 examines the craft of editing. As the alteration of sounds after their initial recording, sequencing, or sound design, editing is the primary axis of producing electronic music dispersed along every stage of making a track. Building upon online fan discussion about the editing prowess of producer David Tipper (Tipper), I consider editing from several perspectives: the zoomed-in view in a DAW, as a meta-tool, as performance, as a way to develop phrasing, and as a balancing of repetition with variation. I conclude by elaborating on producer Mr. Bill's idea of editing as a game of amounts.

Chapter 8 examines arranging and mixing, which comprise the organization of a track into its final linear sequence and sonic balance. In Part 1, I build on two examples drawn from producer-fan discussions on Reddit and Twitter to explore contrasting approaches to arranging: timed/measured sections versus organized tapestries of parts. Next I consider the cons and pros of viewing the arrangement as a linear timeline in the DAW, producer Kara-Lis Coverdale's idea of arrangements as expressions of narrative and complexity, and producer TJ Hertz (Objekt)'s idea of arrangements as by-products of effective automation of parts, transitions, and effects fills. The section concludes with a case study on Autechre's "bladelores", a 2013 track that exemplifies arrangement as an ever-changing process unto itself. In Part 2, I consider mixing in production as a kind of perceptual conjuring that creates enchantment. I present a mixing checklist for hearing the music's unfolding as a sonic landscape that balances four dimensions: stereo field, frequency spectrum, front to back depth, and movement of these dimensions over time. Next I consider lessons from mixing on headphones, the difficulty of balancing "objective" listening with subjective experience, the power of muting and reducing sounds, and how a mix compresses the time of its making. I conclude with six general arranging and mixing concepts relevant to any musical context: contrast, compensations, musical lines, surprise, efficiency, and accumulations.

The concluding chapter brings together the book's broad themes into a final mix. I build on producer Amon Tobin's concept of *pushing forward* to describe the urgency of hearing beyond the particularities of our musical systems to learning how to think with, and through, them. I begin with four lessons from the feedback loop of production: make many small errors through experimentation, work on the most problematic thing, take note of what you are doing, and let your changing perception of your work feed back into your process. Next I propose seven principles by which to push forward one's productions: play and capture, develop the simple into the complex, notice and keep going, refine, reduce and arrange, assess quality, and take your time. I conclude by considering types of failure and how to curate authenticity.

I hope that in this book's gleanings on creativity in electronic music production you will find resonant ideas you can use on your own production Quest. Let's begin.

Notes

1 In an essay about advanced competency as a form of magic, the blogger autotranslucence suggests we absorb "the alien mental models" of other artists: "When you are attempting to learn implicit knowledge that by definition you don't understand, it is important to have a bunch of examples in front of you to feed your brain's pattern-recognition systems" (autotranslucence, 2018).
2 Practitioners including architects, chefs, filmmakers, cartoonists, computer programmers, neuroscientists, mathematicians, inventors, and high-wire walkers. Insights abound in Leski (2015), Adria (2010), Catmull (2014), Lynch (2007), Snider (2017), Kocienda (2018), Eagleman and Brandt (2017), du Sautoy (2020), Taleb (2014), Rubik (2020), and Petit (2015).
3 As you experiment, make mistakes, and disrupt yourself, you can also take notes on your production practice to understand what is working and what isn't. Like your production practice, your notes will create their own feedback loop over time to steer you in a direction you want to go. In appendix to his 1959 book, *The Sociological Imagination*, C. Wright Mills explains the value of keeping a journal to capture any "fringe thoughts" to lay the groundwork for "systematic reflection" on one's work (Mills, 2000: 196).

1 Electronic music production, the producer, and the enchanted DAW

Introduction: Four Tet's studio

On Twitter in 2017, the producer Kieran Hebden (Four Tet) shared a photo of his studio, which consisted of a laptop computer, the DAW software Ableton Live, a small MIDI keyboard, an audio interface, and a set of monitors. "This is where I recorded and mixed the album and all the gear I used," he wrote. Hebden's photo and caption quickly became a meme that inspired humorous responses on Twitter from fellow producers who responded in turn with their own photos of fictitious home studios, some of which featured conspicuously dated technologies, such as 1980s mobile phones and video game consoles (Wilson, 2017).

It was unclear from these responses whether or not musicians were teasing Hebden; after all, publicizing one's minimalist studio setup could be interpreted as a humble brag about one's advanced production skills. But perhaps Hebden was simply showing the production community how little one requires to make music "in the box" of software. The photo captured the intersection of a minimum of gear with a quaint domesticity: the entire studio fit onto a small table, and a window behind it framed a forest setting outside. Intended or not, here was a reminder of the compactness and power of a 21st-century electronic music studio.

In this chapter, I work backwards from Hebden's studio to explore how we arrived at our current moment in electronic music production. I begin by tracing a brief history of music production, the figure of the music producer, and three digital technologies—sampling, MIDI, and DAW software—that remain fundamental to electronic music production today. Next, I consider Ableton Live, the popular DAW whose design and capabilities are a paradigmatic example of software as a "laboratory" in which to work with sound. I conclude by explaining the DAW's basic topology and concepts: arrangement and clip views, the mixer, audio recording and MIDI sequencing, signal routing and effects processing, automation, and mixing.

A brief history of music production and the producer

In our era when producers produce on laptop computers, it is easy to forget that the production process was not always so streamlined and accessible. Recording once required bulky equipment and technicians trained to operate it, and over the past hundred years or so, the evolution of record production and its technologies is a chronicle of ever-increasing fidelity and control over sound. In the early 20th century, recordings were always live recordings. Making a record required ensembles of musicians to be positioned around a mechanical recording device that captured their collective sound via a large conical horn and physically inscribed its vibrations onto a rotating wax disc. For a lead singer or instrumental soloist to be heard, they had to be positioned closest to the device so that their sound was recorded louder than the rest of the ensemble. Even with such proximal adjustments, the sound quality of early ragtime and jazz mono recordings, for instance, was an indistinct mid-range mush lacking clarity and depth.[1]

With the development of electronic microphones, signal amplifiers, and recorders in the mid-1920s, the frequency range of recordings greatly improved, and by the 1940s, magnetic tape recording, developed by the AEG company in Germany, introduced another leap in sound fidelity and led to the first stereo recordings. Tape's ease of editing, which had been impossible with disc recording, led to its quick adoption by radio and the music industry in the United States. Within a few years, components of tape-based audio production had evolved, which remain fundamental today, including: overdubbing and editing, "stereo (since the fifties) and front/back placement, refinement of sounds relative to each other by means of equalization, augmentation with effects, and dynamic control" (Burgess, 2014: 112).

Music production is an omnidirectional and comprehensive way of making music. Production, notes audio production historian Richard Burgess, "fuses the composition, arrangement, orchestration, interpretation, improvisations, timbral qualities, and performance or performances into an immutable sonic whole" (Burgess, 2014: 1). The architect of this fusion is the *producer*, a person with "a specific set of musical competencies pertaining to the *production* of a record" (Shelvock, 2020: 3). The producer figure extends back at least to the work of Les Paul, an American guitarist and recording enthusiast who pioneered various tape recording techniques in the 1940s and 50s, such as *sound on sound*, building layered textures using two tape machines, and *varispeed*, using a sound recorded at half speed then playing it back at regular speed to produce a higher pitch (Burgess, 2014: 51). Paul's experimental, one-man band approach created a template for the producer's craft, and his widely adopted production methods became a toolkit with which to explore "a gradational compositional approach" to recording (ibid.: 54). As heard on songs such as "How High The Moon", a 1951 track which featured overdubbed tracks of guitar, bass, and singing, Paul's multi-track techniques, notes Burgess,

> disengaged production from real time, teased apart its component strands, and approached it as an incremental composition in sound, conflating it with the songwriting, arranging, orchestration, performance, and technical elements.
>
> (Burgess, 2014: 51)

Paul's approach to music production as an *incremental composition in sound* is fundamental because it showed in practice what Burgess identifies as "a magnitudinous expansion of agency by which producers could influence the musical and sonic outcome" of a recording (ibid.: 1). In this way, Paul's tracks presaged and influenced the elaborate multi-track recordings of the Beatles, the Beach Boys, and many other 1960s rock and pop bands. Multi-track recording put the producer at the helm of the studio and created new musical possibilities via a mode of production that effectively turned the recording engineer into a mixer who was fast becoming recognized as "a musical creator of a new kind" (Chanan, 1995: 270).

It was during this time that the producer became known as the person in charge of a recording's overall sound. An example is George Martin, the producer at Abbey Road studios who worked with The Beatles. Martin drew on his experience recording and arranging classical music and comedy albums to help The Beatles achieve unusual sounds in the studio. His creative contributions included arranging parts for string quartet, making tape loop collages, and devising unusual miking and sound design techniques (with Abbey Road's engineer, Geoff Emerick). Another producer from this era is Phil Spector, who devised his "wall of sound" aesthetic by recording and overdubbing large bands with multiple instruments doubling parts, and applying large amounts of echo to the result. This lush sound can be heard on The Ronettes' 1963 track, "Be My Baby." Spector described his incremental-based aesthetic as important as the song itself:

> I was looking for a sound . . . so strong that if the material was not the greatest, the sound would carry the record. It was a case of augmenting, augmenting. It all fitted together like a jigsaw.
>
> (Buskin, 2007)

Beginning in the 1960s and paralleling the rise of the music producer, recording sessions took place mostly in acoustically treated studios equipped with microphones for capturing the sounds of musicians, a mixing console for recording them onto multi-track reel to reel tape machines, and an array of hardware signal processing devices (e.g. compressors, plate reverbs, distortion boxes) through which to alter the sounds either during or post-recording. Within the studio space, there were clear demarcations among musicians, sound engineer, and producer: musicians performed the music, while the engineer and producer decided how to best record and mix the performance. But technological tinkering in the studio blurred these distinct musical roles. Consider a few examples: The Beatles' production adventures with Martin for *Sergeant Pepper's Lonely Club Hearts Band*, Joe Meek's boldly pushing compressors "to create pumping and breathing effects" for his instrumental space age pop album with The Tornados, *Telstar* (Cleveland, 2014), pianist Glenn Gould's "post-performance editorial" decisions to tape-splice together his best takes of Bach's *The Well-Tempered Clavier* with engineers at CBC Radio (Gould, 1966: 53), Osbourne Ruddock (King Tubby)'s dub remixes that dismantled reggae tracks "into something else entirely" by using delay and echo effects (Hebdige, 1987: 83), or engineer Ted Macero's cut and paste recording collages for Miles Davis' *In A Silent Way*. Macero's description of his production approach with Miles explains the goal of this technological tinkering, and remains useful advice for producers today: "How can I make this better? It's good, but electronically can we do something to give it more impact?" (Lee, 1997) Producers such as Paul, Martin, Meek, Spector, Gould, Ruddock, and Macero understood how the techniques of music production could deeply shape a music's sound and feel.[2]

The centerpiece of recording studios had always been the mixing console, a device for combining multiple audio signals on separate channels into a single stereo output. In 1981, Solid State Logic (SSL) introduced the first consoles with *total mix recall*, which allowed producers to "perform" their volume and effects changes to mixes in a DJ way and have the console remember these moves through computerized automation, "thus eliminating the reliance on human memory or indeed channel strip notes jotted on pieces of paper" (Bennett, 2019: 38). Crucially, with these consoles the mixing component of music production "began to be an iterative process rather than a performance" (Burgess, 2014: 101). In this way, SSL automation presaged the vast automation capabilities that are now a standard feature of the DAW software on your laptop.

Fast forward back to Hebden's home studio: electronic music producers create in the tradition of record production, but they control and oversee every aspect of a musical project—from conception, performance and recording, sound design and editing, to arranging and mixing. Indeed, it is common practice, notes Richard Burgess, "for the same person . . . to write, perform, arrange, engineer, and produce hit tracks" (Burgess, 2020: 95). Today, electronic music producers mostly work alone using DAW-based musical systems, sometimes with additional hardware such as synthesizers and signal processors. Playing the roles of musician, sound designer, and engineer overseeing every aspect of their tracks, producers record, edit, and mix as they go, shaping and refining the music at each step of the process. In short, in electronic music production, the producer is the composer (Moorefield, 2005).

Recording as the basis for composing

In 1979, the producer and production philosopher Brian Eno gave a talk in New York City titled "The Recording Studio as a Compositional Tool" (Eno, 1979). Amplifying (though not acknowledging) the *studio-as-instrument* approach of producers such as Les Paul and Joe Meek from a few decades earlier, Eno explained the virtues of using the studio's functionalities as a springboard for composing. Eno, who once played synthesizers with Roxy Music, recognized the potential of the studio not merely as a place for recording *already composed* music, but for *recording as the basis for composing*. Rather than rehearse and arrange music first and then go into the studio to record a completed track, Eno advocated for tape recording-based *composing music in an additive way*—for example, by overdubbing, sound designing, and editing the results—and being responsive to the chance accidents and discoveries that inevitably arise while working among the studio's connected technologies. In a 2011 interview, he explained the approach:

> You could make a piece over an extended period of time—it didn't have to preexist the process, you could make it up as you went. And you could make it like you would a painting—you could put something on, scrape something else off. It stopped being something that was located at one moment in time. It started being a process that you could engage in over months, or even years.
>
> (Baccigaluppi & Crane, 2011)

Eno had been putting this philosophy into play since the mid-1970s using his Oblique Strategies, a set of aphorisms he devised with the visual artist Peter Schmidt that were printed on a deck of cards and intended to help artists think differently about creative problems. Eno had used these aphorisms and the studio as a compositional tool approach in his work with the ambient keyboardist Harold Budd, King Crimson guitarist Robert Fripp, and later with David Byrne, U2, and Coldplay. In these collaborations, a recurring theme in Eno's producing was not so much pursuing a particular sound, but rather encouraging artists to focus on the potentials of the process at hand, and most importantly, maintain a supple awareness in the studio. For Eno, the most important idea is *attention*:

> I've come to think that attention is the most important thing in a studio situation. The attention to notice when something new is starting, the attention to pick up on the mood in the room and not be emotionally clumsy, the attention to see what's needed before it is actually needed, the attention that arises from staying awake while you're working instead of lapsing into autopilot.
>
> (Baccigaluppi & Crane, 2011)

MIDI

In the early 1980s, the processes of popular music production and the studio's configuration as a compositional tool expanded exponentially in response to two technological developments. The first was an array of relatively affordable synthesizers, sequencers, drum machines, and samplers from companies including Roland, Korg, Yamaha, Oberheim, Sequential Circuits, Linn Electronics, Ensoniq, E-mu Systems, and Akai. The second development was an electronics communication protocol called MIDI, or *musical instrument digital interface*. Jointly devised by Roland and Sequential Circuits' founder Dave Smith, MIDI enabled electronic instruments to be connected and synchronized with one another. MIDI remains essential to electronic music production today. Smith explains that the reason MIDI has endured is simply that it was cheap to implement, and "still covers about 95% of what people need to do" (Schmidt, 2014). MIDI and popular MIDI-compatible instruments like the Yamaha DX-7 synthesizer, notes Ryan Didduck, "helped reorganize the structure of musical data . . . altering the aesthetic qualities of music subsequently made with them" (Didduck, 2017: 90). Being able to perfectly sync MIDI devices was as fundamental to the aesthetics of 1980s music production as Les Paul's incremental composition in sound technique had been to the making of classic rock records in the 1960s. With a MIDI-connected synthesizer and drum machine, for example, a musician could trigger multiple patterns that played together *perfectly*. Repeating patterns of MIDI notes known as *sequences* could be recorded and stored onboard the instruments themselves, and soon, in MIDI sequencing software such as Cubase (discussed further in this chapter). In short, MIDI opened up a new way of thinking about composing in terms of "an unprecedented degree of post-performance control to an instrumentalist or engineer" (Barry, 2017: 158).

From its inception, MIDI had its detractors who said that the protocol constrained the sound of the music made with it by imposing a binary/on-off/one-zero grid. MIDI's grid, notes Robert Barry, suggested "certain ways of working—the use of quantized rhythms and discrete, tempered pitch classes—by making them simpler to do, making them the 'default' settings" (Barry, 2017: 159). Yet it was precisely MIDI's default constraints that freed music producers from the instrumental conventions of a traditional band of musicians, a topic we discuss further in Chapter 3. MIDI shaped popular music's sound by nudging musicians to recognize a new composing toolkit comprised of programmed drum patterns, step-sequenced melodies, and chord progressions with which to work. MIDI's fixed grid, observes electronic music producer Stefan Goldmann, became a structural feature, shifting producers' attention onto a new field of play:

> MIDI sequencing and looped audio hardly allowed for the micro rhythmic and dynamic variations a manual performance granted, yet any detail that was mere nuance before could be fixed rigidly and thus become structurally relevant. It is precisely the repetitive grid that can make minute variations as stable as pitches on a piano. New fields of play. This very shifting of attention, the reallocation of both creativity and generic aspects is what keeps music diverse, constantly evolving without surpassing its predecessors.
>
> (Goldmann, 2015: 18)

Digital sampling

A few years prior to MIDI, digital sampling was developed by Kim Ryrie and Peter Vogel of the Australian digital audio company Fairlight, whose Fairlight instrument, released in 1979, combined synthesis and digital sampling into a kind of pre-DAW music workstation. "The digital memory recorder it contained was an important feature," notes Paul Harkins in his history of digital sampling, "anticipating the development of MIDI sequencing and the Digital Audio Workstation (DAW)" (Harkins, 2020: 6-7). Although the concept and practice extends back to the 1940s with Pierre Schaeffer's *musique concrète* tape loops of train sounds, collage experiments that "echo still in all contemporary sample-based music" (Brend, 2012: 24), Ryrie and Vogel coined the term *sampling* to describe their process of digitally playing back short segments of an audio recording at different pitches to reconstitute its sound. As Vogel summed up the seemingly magical workings of their invention: "Just take the sounds, whack them in the memory and away you go" (Hamer, 2005). The Fairlight was used by experimentally minded pop producers, including Trevor Horn, Kate Bush, and Peter Gabriel, who wove samples of everyday sounds and non-western instruments into their tracks (Harkins, 2020: 28–32). "Most other people didn't understand at the time," said Horn, "sampling was like a mystical world" (Peel, 2005). Another early sampling system was the Synclavier I and Synclavier II by New England Digital. Released in 1977 and 1980, both of these instruments were marketed as "Tapeless Studios." While the Fairlight and Synclavier cost tens of thousands of dollars and were used mainly by wealthy pop stars, the instruments were harbingers of future ways of working with sound as a malleable material.

By the mid-1980s, more affordable digital samplers, such Ensoniq's Mirage, E-mu Systems' SP-12 and SP-1200, and Akai's MPC, were helping musicians devise new production workflows for building tracks. Hip hop musicians, for example, moved from using turntables to samplers, using the SP-12 /1200 and the MPC60 to create breakbeat- and sample-based song structures. Samplers were becoming recognized as meta-instruments built from already recorded sounds that "made sonic material much more reusable, malleable and open to transformation," thereby fundamentally changing the production process (Stratchan, 2017: 5). Here is drum and bass producer Roni Size describing how he and his Reprazent production crew made their 1997 track, "Brown Paper Bag":

> We wanted to make music that sounded like the future. . . 'Brown Paper Bag' started off as samples of double bass licks. I chopped them up on the sampler and suddenly there was a song.
>
> (Simpson, 2018a)

Throughout the 1980s, 90s, and beyond, future-oriented producers used samplers and other MIDI-connected electronic instruments to build tracks that would define the sound of new wave pop, hip hop, and the myriad varieties of electronic dance music.

Reams of musical data in accessible form: Steinberg's Cubase

Along with MIDI and affordable digital samplers, the 1980s saw the emergence of first-generation DAW software. Computers had been used in music since the 1950s. In 1951, the Ferranti Mark 1 computer at the University of Manchester was programmed to play "God Save The Queen" and in 1957, Max Matthews used his MUSIC I software at Bell Labs to perform a 17-second composition (Prior, 2018: 63). By the 1970s and early 80s, computer processing had improved and the first home computers, such as the Apple II and Atari 520ST, appeared. As Michael Chanan notes in his history of the composer, the computer during this time was fast becoming "the universal electronic musical instrument, subsuming all types of electronic and electro-acoustic music production" (Chanan, 1999: 272). The emergence of DAW software, then, brought together a constellation of technologies. In his book *Popular Music, Digital Technology and Society*, Nick Prior locates DAWs as "fitted neatly into an historical moment that drew together pre-existing innovations in keyboard-controlled performance programs . . . MIDI . . . the lowering of manufacturing costs and the increasing domestication of the PC" (Prior, 2018: 71).

An early DAW innovator was Steinberg Research, a company formed in 1984 by Karl "Charlie" Steinberg, a musician and studio engineer, and Manfred Rürup, a keyboard player. Steinberg's first product was a 16-track MIDI sequencer called Multitrack Recorder, which was soon developed into Pro-16, a 16-track MIDI sequencer for the Commodore 64, a popular home computer at the time. Inspired by an Atari 520ST they had seen at a trade show that had built-in MIDI ports and a mouse-controlled interface, Steinberg and Rürup saw "the possibility of making a professional product that could be used everywhere without additional interfaces" (Ramage, 2010). In 1983, Steinberg released the successor to Pro-16, Pro-24, designed for Atari's 520ST, a more powerful computer than the Commodore. Pro-24 offered 24 tracks of MIDI sequencing, each track with independent controls for volume, panning, and note velocity. The software also introduced several editing screens, including a MIDI edit list and a step sequencer-style grid editor for drum parts, quantization timing correction, and score notation. Most importantly, it was the first software to feature an arrangement page that "listed the tracks vertically and the timeline horizontally, something which actually changed the way people think about composing" (Twells, 2016). In 1989, Steinberg released Cubit, soon renamed Cubase. A magazine ad touted Steinberg's Visual Song Processing (ViSP) technology, which was a way to describe the horizontal display of a song's MIDI parts:

> With ViSP you have a clear visual representation of your song at all times, you can watch as your music takes shape from beginning to end and you have total accessibility, without constraint, to go anywhere, anytime, in real-time.
>
> (RetroSynthAds, 2019)

Displaying a song's arrangement of MIDI parts was just one of the DAW's many conveniences. A 1989 review of Cubase in *Music Technology* magazine described the software as "the ultimate expression of the programmer's art" and declared, "the onus is on the machine to come to terms with the often capricious temperament of its human operator rather than vice versa" (Lord, 1989). Curiously, the reviewer, Nigel Lord, got the relationship between the DAW and its user backwards: over the subsequent decades the onus would be on electronic music producers to figure out ways of working around the quirks of computer-based production. Nevertheless, Lord's description of Cubase's capabilities for recording, editing, and arranging MIDI still resonates. The software, he said, "leaves you feeling slightly breathless is in its ability to see things quite literally—your way. It can lay before you reams of fiendishly complex data in an astonishingly accessible—and above all, human form" (ibid.). In his study of synth pop producer Trevor Horn, Timothy Warner explains in broad strokes how MIDI sequencing's *reams of data* impacted 1980s electronic music production practices, recalling Eno's *composing in an additive way* concept:

> The development of MIDI, and particularly MIDI sequencing programs, during the 1980s enabled the manipulation of musical material in new ways: ideas can be developed by working directly with sound and then copied, 'pasted' and, if necessary, modified wherever might seem appropriate. Hence, *rather than beginning with a specific structure in mind, musicians can allow structures to evolve though an investigation of the potential for combining the various ideas.*
>
> (Warner, 2003: 87, italics added)

The first version of Cubase for the Apple Computer was released in 1990, and in 1991 Cubase Audio was released, which supported recording audio tracks as well as MIDI. In 1996, Cubase introduced VST (Virtual Studio Technology), a protocol for virtual effects *plug-ins* to be used inside the software, allowing Cubase "to use the host computer's native processing power to manipulate audio in real time" (Future Music, 2011). In 1999, Steinberg's VST 2.0 protocol allowed plug-ins to be controlled by MIDI. This functionality has had perhaps as profound an impact on electronic music production as MIDI itself. VST 2.0 allowed producers to use MIDI-mapped parameter automation in their productions—in effect, bringing the total recall of an SSL mixing console into new sonic territories. Steinberg soon licensed the VST technology to other third party plug-in developers, which in turn led to VST instruments and effects becoming a leading plug-in format. Finally, Cubase's graphical user interface (GUI) introduced features that have since become staples of all DAWs: it displayed the MIDI data on each track and allowed users to move MIDI sequences around by clicking on and dragging them using their computer mouse, while right-clicking brought up a toolbox of editing options (Figure 1.1).

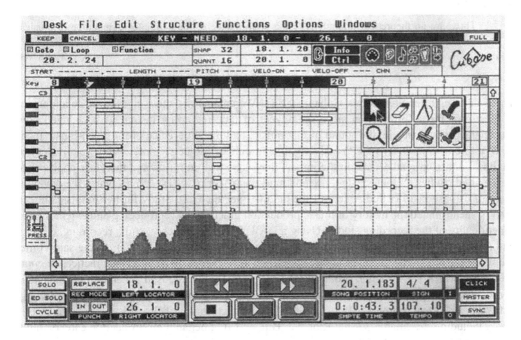

Figure 1.1 Key Edit window in Steinberg's Cubase DAW, c. 1989

In the 1990s, as Steinberg was developing new versions of Cubase, other music technology companies developed their own DAW software. In 1990, Mark of the Unicorn released Digital Performer; in 1991, Digidesign launched ProTools; and in 1992, Emagic created Emagic Logic (which became Apple's Logic Pro). These programs were soon joined by Image-Line's Fruity Loops in 1997, Propellerhead's Reason in 2000, and Ableton's Live in 2001.

These DAWs looked visually similar, providing an Arrangement page in which to display musical data on a horizontal, left-to-right timeline, coupled with a mixing console screen with track volume faders, effects Sends, Pan knobs, Mute and Solo buttons, and other controls. With their timelines, pop-up windows, database-style layouts, drop-down menus (Stratchan, 2017: 77), drag-drop, cut-paste, and other audio editing functions (Holmes, 2008: 301), DAWs re-framed music production as "a malleable digital landscape" (Prior, 2009: 87) in which the graphic representation of musical events presented musicians with "sound as an object rather than a stream" (Zagorski-Thomas, 2014: 134). Users of DAW software had to learn a new production workflow, specifically: they "were expected to use the computer mouse to select and change parameters" (Prior, 2008). The DAW, explains Nick Prior, required the producer learn to *think digitally*:

> Thinking digitally . . . requires a shift in the attachments, modes and haptic efforts needed to compose within techno-spaces comprising windows-type arrangements, menus, scroll bars and cursors that are controlled with a mouse, trackpad or MIDI controller.
>
> (Prior, 2018: 74)

All of this to say that the DAW + computers irrevocably introduced a new way of music production: mouse pointing-and-clicking.

The DAW

> There are no theoretical limitations to the performance of the computer as a source of musical sounds, in contrast to the performance of ordinary instruments.
>
> —Max V. Matthews[3]

Three decades after Steinberg's Cubase, the DAW remains the most significant development in electronic music production since the advent of tape recording insofar as it changed how musicians produce music, the sound of that music, and the experience of musician-music technology interaction. DAWs are "surely significant cultural objects," notes Prior, "equivalent to studios, guitars or synthesizers in the impact they have had on music and therefore appropriate for detailed studies on their design, use and marketing" (Prior, 2018: 75). Between the development of Cubase in 1989 and Ableton Live in 2001 (see the Case Study further in this chapter), DAWs encapsulated the history of music production practices by reproducing and extending the functionality of the recording studio. Using a DAW, MIDI sequencing, audio recording, editing, and mixing "could now be done in one environment" (Stratchan, 2017: 77). At the same time, the scope the electronic music producer broadened as his/her skills merged the practices of composer, performer, DJ, arranger, producer, and engineer. By the 2000s, a generation of musicians "with little or no experience of the hardware studio" were learning production in exclusively software-based environments (Prior, 2008: 924). In fact, many of the artists discussed in this book, including deadmau5, Virtual Riot, Mr. Bill, and Jlin, began their production careers in the DAWs Cubase, Reason, FL Studio, and Live.

Like a microphone that picks up sounds from all over, the DAW is an omnidirectional compositional interface, inviting the producer into music making from multiple vantage points. Where a band once composed and rehearsed a song and then went into a studio to record it, now the production moment is fluid, connecting what were previously separate activities and domains of musical or technical expertise into a single compositional experience. Moreover, a significant, yet perhaps under remarked fact of the DAW is its broad accessibility. Using a DAW, a novice musician can produce and learn about production techniques by recording audio, sequencing MIDI, looping and arranging beats and chords, and mixing the results. Expensive hardware that was once found only in professional studios (e.g. the Teletronix LA-2A compressor) or university electronic music labs (e.g. the Buchla 100 Series modular synthesizer) is convincingly emulated by affordable VST plug-ins inside the DAW. As Matthew Shelvock notes in his study of hip hop production, "cumbersome analog processes which required bulky hardware" are now digitally mimicked (Shelvock, 2020: 2). The DAW, summarizes Richard Burgess, "revolutionized the art of music production. It bestowed upon many more producers, songwriters, and artists than ever before the power to manipulate and optimize music, as easily they can words with a word processor" (Burgess, 2014: 134). In sum, a DAW encapsulates a plethora of production tools, awaiting our creativity. Let us turn now to some of the DAW's potentials and functionalities.

A laboratory for thinking

Building on his notion of the studio as a composing instrument unto itself, Brian Eno encouraged musicians to treat it "as a laboratory for conceptual thinking—rather than as a mere tool." Eno's suggestion to think of the studio as a compositional tool applies equally to the DAW. Producers use the software as a space in which to "work up their musical experiments" (Hennion, 1989: 406)—that is, to try out sounds, to apply concepts (e.g. *What would it sound like if I played this audio at half speed?*), make mistakes and iterate to accelerate their learning, and save each idea they try (e.g. a sample, a sequence, an effects rack) in case they want to revisit it later. As a laboratory in which we can both hear and see our sounds, the DAW invites ongoing experimentation with ways of making, modifying, and combining sounds. This experimentation can be overwhelming to contemplate: there are no surefire methods by which to produce, and every producer evolves his or her own ever-changing methods and routes discovered over the course of their production Quests. However, there are some fundamental concepts of DAW-based electronic music production. We can apply these concepts to our work, no matter what kind of music we make.

Taking creative cues from software

The DAW has numerous basic production techniques built into its functionalities, awaiting you to put them into action. So rather than ponder what ideas you may or may not have (which is a common source of "creative blocks"), use the software's design as a foundation and springboard for your creativity. Take a moment to consider: every function, parameter, setting, preset, or control in the DAW is a tool you can use. Each of the DAW's functionalities and capabilities—audio recording, sampling, MIDI sequencing and part layering, signal and effects processing, editing, arranging, and mixing—contains a plethora of production techniques to explore.[4] Producer Boris Brejcha reminds us, for example, how every parameter on a VST synthesizer or effect plug-in can be adjusted and automated:

> You can do automation with every knob you have on a synthesizer. And this makes your track—it starts to live, because the sound is moving a lot, it's changing a lot, and so it sounds not all the time the same. I think this is the best choice: playing more with effects, for example, or with automations than using more sounds.
>
> (Brejcha, 2018)

In sum, the DAW's power is that it plays multiple roles:

- it is a shape-shifter, capable of modeling other sound sources, spaces, or processes—a studio, a collection of instruments, a sequencer, a score, an algorithmic process, a mixing console, a chain of signal and effect processors—which the producer can deploy to play with audio and MIDI, editing, arrangements, and mixes
- it amplifies ideas, positioning the producer to be analytical, architectural, and precise about the intuitive, gestural, and vague contours of the creative process.

Case study: Ableton Live

Ableton Live was developed in 2001 by two electronic musicians, Gerhard Behles and Robert Henke, who designed the software to seamlessly integrate composing, production, and performance. Similar to, yet unlike, longstanding DAW software like Cubase or Logic, whose staid visual interfaces are modeled on the look of analog studio mixing consoles and other hardware, Live avoided skeuomorphic representations (Lagomarsino, 2017), conceived instead as a sui generis "technoscape" with a flat and minimalist GUI (Prior, 2008: 923). Live was unique in two ways: it was the first software that could seamlessly loop and time-warp audio samples to synchronize to a track's tempo, and its Session view offered a new way of visualizing and organizing musical data as building blocks called *clips*. In Live, the clip is "a piece of musical material: a melody, a drum pattern, a bassline or a complete song" that plays once or loops, and groups of clips can be arranged into horizontal *scenes* triggered collectively or in various combinations "to create larger musical structures" (Ableton, 2018: 4). For example, a track with three clips might have a 1-bar kick drum pattern, a 20-bar chord sequence, and a 4-minute field recording. Triggering these clips to play together in a scene will create ever-changing combinations of sounds as the clips loop at different points.

As distinctive as Live appeared, its use of clips as the essential "unit of creativity" (Zagorski-Thomas, 2014: 76) as well as its "naturalization of the loop paradigm/sampling aesthetic within the DAW" distilled key moments in electronic music's history (Marrington, 2019: 69n2). Using the software, it is not especially difficult to create *musique concrète*-style sound collages, canonic loops, or remixes. The software's potential uses are amplified by Ableton's online creative ecosystem that positions Live as a part of the experimentally-oriented producer's creative toolkit. Ableton's ecosystem includes numerous video tutorials and interviews with Live users about their production techniques, and Loop, the company's annual conference on music, creativity, and technology. The design and functionalities of Live have made it among the most impactful music production technologies of the past two decades, earning a loyal following among electronic music producers.[5] Many musicians (including this author) have embraced Live's non-linear, modular, and recombinant musical aesthetic, building tracks by experimenting with various combinations of looping clips before recording them into a linear arrangement.

DAW topology and concepts

Minutiae on a screen: arrangement and "clip" views

The DAW represents sounds as visual objects on the screen, which means that whether you work with audio samples or MIDI sequences, when you use a DAW *you look at your sounds while listening to them* (or, listen while looking). The producer Jon Hopkins describes the "accepted fact" of the DAW's visuality in electronic music production, which assists the producer's "infinitesimal tweaking across loads of parameters" (Smith, 2018b). "A good 90% of the process", Hopkins admits of his own production work, "is essentially just looking at the screen and playing around with things" (Rancic, 2018). Consider then, the screen as a window onto the inner workings of your production project, a map of its terrain. As your monitors or headphones reproduce the music's acoustic space, the screen displays a schematic in which all of the sonic components of your track are connected, compressed into a "single entity" (Marrington, 2011). All DAWs have an Arrangement view page, where the music is laid out in a linear, left to right timeline, as a series of audio or MIDI events (Constantinou, 2019: 235). Live and, more recently, other DAWs such as Apple's Logic Pro and Mark Of The Unicorn's Digital Performer, include non-linear clip displays of audio and MIDI events that can be looped and triggered in any combination (Figure 1.2).

The mixer

All DAWs include a mixing console page, where each audio and MIDI track is displayed as a vertical channel strip with input/output routing options, a volume fader, a pan control, mute, solo and record buttons, effects Inserts, and effects Send controls. Mixing is generally done on this page, although mixing automation is drawn in on the project's Arrangement page.

Audio recording and MIDI sequencing

The primary purpose of a DAW is to record audio and MIDI data. A project is comprised of any combination of audio and/or MIDI tracks. Sounds for audio tracks can come from audio samples or from a live performance captured by microphone or direct input and sent into the computer

Figure 1.2 Ableton Live's Session view

via an audio interface. Sounds for MIDI tracks can come from the DAW's own VST instruments, from third-party VSTs, or from external hardware sources routed into the DAW.

Signal routing and effects processing

Within the DAW, sounds can be routed in many ways. A fundamental way of routing MIDI and audio is through signal and effects processors such as compressors, reverbs, delays, etc. Such signal and effects processing can be inserted directly onto an audio or MIDI track via an *Insert*, or placed onto an effects *Return* track controlled by a track's Send control. For example, a reverb inserted onto a Return track can be sent (routed) in varying amounts to multiple audio and MIDI tracks simultaneously. This processing technique is commonly used to create a sense that different sonic elements in a mix inhabit the same acoustic space. Return tracks are also a CPU-friendly way of applying effects. In the DAW, signal routing and effects have no end, limited only by the producer's ideas about where to send sounds and how to treat them.

Automation

Automation refers to directing the DAW to perform specific production tasks over time. The most common type of automation is *mix automation*—for instance, having the volume of an individual track smoothly increase or decrease over a portion of the arrangement. However, automation can be applied to virtually every parameter within the DAW and its VST instruments and effects. For example, one might automate a reverb's Size setting to change from a small to a large space, a distortion effect from soft to heavy, or a granular sampler from subtle to glitched. In general, automation is a way to add fluid and sometimes imperceptible changes to a track's parts, and therefore a sense of life to a production. Producers use automation to shape dynamics and add contrast, change which sounds are "in focus," and morph parameters of audio samples and synthesized sounds to create textural-timbral shifts impossible to achieve any other way. As with signal routing and effecting processing, there is no end to the producer's automation options.

Mixing

When a producer creates a track, s/he is designing a circuit in which a collection of sounds co-exist and interact together in time. A mix can be as simple as a balancing the volumes of a beat and a bass line, or as complex as blending dozens of sonic textures into an intricate, evolving whole. Mixing is the process of combining signals from different audio and MIDI tracks to create a composite stereo blend. Blending tracks together typically involves balancing their individual volume levels, using equalization (EQ) to shape them, and applying signal processing (e.g. compression) and effects processing (e.g. reverb) to create a sense of the music as a living and dimensional space. We explore some general mixing concepts in Chapter 7.

Conclusion

This chapter has traced a brief history of music production, focusing on the role of the producer, the development of MIDI, digital sampling, and DAW software, and provided a brief overview of DAW topology and concepts. One broad theme that emerges over the history of record production is that as technologies have evolved and gone digital, so too has the electronic music producer, whose responsibilities now include overseeing every stage of a music's production inside the DAW. Techniques such as using an incremental/additive approach to recording, sequencing, and sample manipulation form the toolkit of most every producer who organizes and edits reams of musical data in the laboratory that is their DAW. Producers are attuned to seeing and thinking about tracks as malleable arrangements of MIDI and audio objects, and it is an everyday production practice to discover new sounds and take creative cues by experimenting with a music software's functionalities, such as combining clips in Live's Session view, for example. In sum, let us return to Four Tet's small laptop-based set up with which he recorded and mixed his album. We might reflect on the fact that his studio is more than a reminder of the power of 21st-century electronic production tools. With a DAW and various VST instrument and effects plug-ins, Hebden uses equipment available to *any* producer. It might not look like much. But of course, the photo leaves out the considerable tacit knowledge, experience, and good taste that informs Hebden's producing. It leaves out a sense of how a producer uses a *musical system*, a topic to which we turn next.

Notes

1 For a vividly detailed history of recording and recorded music, see Milner (2010).
2 Two excellent overviews of producers' use of the recording studio, from the 1950s to the present, are Moorefield (2005) and Bell (2018).
3 (M. V. Matthews, 1963).
4 This idea was articulated in 1975 by Jon Appleton and Ronald Perera in *The Development and Practice of Electronic Music*. "Each parameter is associated with the devices and techniques of the tape studio," they note, "and suggest to the composer distinct creative possibilities" (Appleton & Perera, 1975: 70).
5 One metric of Live's popularity: at the time of this writing, the software's subreddit ("r/ableton") has 175,000 subscribers, compared to 8,000 for Cubase ("r/cubase").

2 Musical systems and production workflows

Introduction: musical systems and production workflows

No matter what your level of experience, you can set up your own *musical system* through which to produce. A musical system is the configuration of your electronic music production toolkit. For many producers, this is an ever-changing assemblage of software (e.g. a DAW and VST instrument and effects plug-ins), hardware (e.g. modular synthesizers, drum machines), or a combination of the two (Butler, 2014: 93). No matter what kind of equipment you work with, it is essential that you configure—and continue to re-configure—your own production system because this system shapes how you work. "I try to expand the approach all the time," says producer Steven Ellison (Flying Lotus), "by using different software or adopting different tools for the production" (Nagshineh, 2013). In this chapter I explore musical systems and workflows for interacting with them. I begin by explaining how musical systems are kinds of complex systems whose nonlinearities steer the producer, providing ideas upon which to build. I then explore numerous examples of production systems, with a case study on producer Brian Eno's sonic treatments and generative loop machines. The chapter concludes with suggestions for setting up a system through preferred ways of working, controllers, templates and sound sets, and how to begin the production process.

The importance of musical systems was observed 50 years ago by Allen Strange, in his 1971 book about the methods and technologies of electronic music composition, *Electronic Music: Systems, Techniques, and Controls*. Although Strange was writing about analog technologies of the time such as voltage-controlled synthesizers, he anticipated the creative challenges of digital music production when he defined musical systems as structures that set up yet to be heard possibilities:

> The contemporary electronic music system has no pre-defined structure, but is initially a collection of possibilities—a set of musical variables or parameters such as pitch, loudness, space, timbre, etc., that exist in an undedicated state. The contemporary electronic music system offers several potential means for pitch control; it has capabilities of shaping articulation and loudness; it provides many methods for the control of timbre and density, and so on. All of these possible sources and controls are components of a yet to be produced musical event.
>
> (Strange, 1971: 3)

More recently, Strange's insight is picked up by composer and theorist Francois J. Bonnet in his manifesto, *The Music To Come*. Bonnet's observation of how music creation requires setting up "the conditions of possibility" speaks to the importance of musical systems in electronic music production:

> The challenge of any musical creation must be to invent the conditions of possibility for an appearing of the musical, but nothing else, because that is all that lies within its power. You don't create music; you create environments conducive to the advent of music.
>
> (Bonnet, 2020: 30)

Building upon Strange and Bonnet's notions of musical systems as structuring conditions for musical possibility, consider the idea even more holistically: imagine the musical system extending *beyond* the configuration of the equipment we use. Such a system would also include the musics we listen to, how we approach and generate musical ideas, how we think about our work, and our general mindset. *The basis for your music production is who you are, and how you are.* This fact is humorously illustrated in a video interview by Telekom Electronic Beats with the Belgian techno producer Peter van Hoesen in which Hoesen explains how he uses an extensive collection of modular synthesizers and other equipment to make music. In the video's comments, an astute producer-fan named azmotronik reminds us of the *real* basis of Hoesen's musical system: "Actually, the most important piece of the equipment in this studio is PVH" (Hoesen, 2017). Even a seemingly simple musical system then, is in fact complex in that it connects the technological with the biological, equipment with its user. Here, for example, is producer r beny connecting the two realms, describing his system as "searching for textures or sounds I like, inspired by whatever I'm inspired by in the moment . . . emotions, other music, people, places, things" (Stationary Travels, 2018).

A *workflow* is how you use your musical system by creatively interacting with it through a sequence of decision-making, tinkering, and technical moves. As Nick Prior notes, the configuration of a musician's system favors "particular kinds of creative processes" by which to use it (Prior, 2009: 82). At the same time, there are virtually endless ways to work, and by configuring your system's components variously, your workflows help you discover paths of musical expression.[1] Think of workflow, then, as your ways of producing music—which includes sound designing, improvising and recording, sequencing, editing, arranging, and mixing. Through your workflow, your work flows. "The goal," says producer Jason Chung (Nosaj Thing), "is to be able to output your ideas without thinking at all" (Future Music, 2018b).

Musical systems as complex systems

It can be useful to think about electronic music production as a kind of *complex* system. A complex system is any system composed of many components that interact with one another to create unpredictable collective behavior. Examples include the organization of human cells, traffic patterns, global climate, and viruses. The behavior of such systems is difficult to model due to the way their components interact with one another in unforeseen ways. Complex systems frequently function in a state referred to as the "edge of chaos" to produce order and patterns from "intrinsically noisy" components (Ziemelis & Allen, 2001). Moreover, since there are dependencies among a system's components, new components added can cause the system to behave in unpredictable ways. Complex systems like traffic patterns and global climate are typically nonlinear and have built-in feedback loops. This leads the systems to exhibit a property known as *emergence*, a condition in which "the action of the whole is more than the sum of the actions of the parts" (Holland, 2014: 2).

In her 1972 book, *An Individual Note*, pioneering electronic musician Daphne Oram reached towards complexity's dynamics when she compared electronic music composing to steering a yacht in a storm:

> The resulting flow is a complex pattern of tensions and relaxations which evolve as the musical material is worked out. The words "controlled" and "worked out" do not really convey what I mean. There seem to be no suitable English words. I am hunting for some word which brings a hint of the skillful yachtsman in fierce mid-Atlantic, guiding and controlling his craft and yet being taken along with it, sensing the best way to manage his vessel, freely changing his mind as unforeseen circumstances evolve, yet always applying the greatest discipline to himself and his seamanship. . . . The composer has to guide and evolve his material in all its aspects.
>
> (Oram, 2016: 27)

With this in mind, consider electronic music production as an analogous kind of complex system capable of generating unpredictable and fascinating cascades of sound with which you can interact, derive ideas from, partially steer, and feed back into your tracks. In production, unpredictable complexities—what producers usually call "happy accidents"—are always our allies. In "Machines and Human Creativity," the conclusion to his 1980 book, *Introduction To Computer Music*, Wayne Bateman speculated on the conditions necessary for what he calls "creative incidents" to transpire in computer-based music. Like Oram, Bateman hinted at complex systems when describing how artistic works are produced as one interacts with accidental discoveries unpredictably generated by some kind of system:

> the evolution of artistic works . . . is a continuum of spontaneous, creative incidents. The incidents seem to occur much more stochastically than deterministically. They happen unpredictably and by apparent accident, but also by improbable coincidence.
>
> (Bateman, 1980: 247)

This idea of music production being a complex system, then, is not new, although today it is somewhat obliquely referred to in the discourse of electronic music producers themselves. Most producers do not talk about it explicitly. Instead, they tacitly engage with it through their workflows. But some do talk about complexity in one way or another. For example, the idea is explained by Mr. Bill (Bill Day), a producer and well-known online educator who uses the phrase *reactive artistry* to describe how he approaches his DAW, Ableton Live. In interviews and tutorials, Day mentions a complexity-generating technique for avoiding creative blocks: whenever he is unsure what to do next while working on a track, he tries out new ways of processing sound. Day compares this processing to a Rube Goldberg machine that generates ideas despite its user's own uncertainty, giving the producer something to react to:

> What I react to is just things that I make the computer do. So I set up some kind of sound design system—like a giant chain of effects that acts like a Rube Goldberg machine: maybe you'll have a one-shot [sample] that goes into a reverb, and then that reverb gets sent through some wave shaper, and the wave shaper glitches it out, and then the glitchy stuff gets re-arranged by some weird plug-in. I'll set up something like that and then just run tons of shit through it and see what comes out. . . . In that sense, I don't even believe that I had the idea. I just set a thing up, press the button, and then it gave me the idea. . . . I don't think creativity is this sort of one way street.
>
> (Gee, 2019).[2]

Producer Tim Hecker explains how he connects a mixing console, guitar pedals, and other hardware effects with a collection of software patches for Max (a programming environment by Cycling'74) as a musical system through which to route sounds and produce unpredictable results. Referring to it as a "data flow as a stream you can capture and mold on the fly," Hecker elaborates on his system:

> they all connect in a network so you can wire all these things together, and have them all modulate each other, and talk to each other. It's kind of like modular synthesis but it's computer [-based], so you can just work with digital audio in a really fluid way. Treating it like a river that flows between all these different things. It's just a lot more flexible and open sourced. . . . I just route things to each other and do feedback loops.
>
> (Burns, 2016)

Another example: in a YouTube tutorial about Pigments, a VST synthesizer by the software company Arturia, producer Das Glitch describes a process similar to that of Mr. Bill, configuring the synthesizer's sequencer settings to auto-generate randomizations on pitch, octave, velocity, and other parameters. The GUI of Pigments' sequencer includes "random" knobs, represented by small dice icons (Figure 2.1). Glitch explains how he uses this randomization to guide his workflow:

> You can actually choose when to regenerate the different randoms . . . It's insane what you can do here in terms of experimenting to come up with a new idea, then write it to the preset, and then the idea is there. For me, I would just keep generating these randoms until something cool is automatically spat at me, and then go with that.
>
> (Glitch, 2019)

Producer TJ Hertz (Objekt), a producer we will hear from several times throughout this book, goes one step further. Hertz understands his production system as doing the musical imagining on his behalf. He allows the processes of his system to generate difficult-to-predict sonic complexities with which to work:

> My approach to sound is more about letting processes and equipment do the imagination for me, and arranging the results of that myself . . . A lot of the time it has to be something with

Figure 2.1 Sequencer for Arturia's Pigments VST, with its small dice icons for randomization

an element of unpredictability or uncontrollability . . . the results of which I can sculpt or steer into the right direction rather than create the humanism to begin with.

(Martin, 2018)

Kaitlyn Aurelia Smith, who produces using modular synthesizers, approaches her musical system as a kind of orchestra, conceptualizing its sounds as instrumental sections. She sees her creative role as both composer and conductor:

My favorite way to approach composition on the modular synthesizers is to imagine that I'm a conductor. I like to set up all the voices and get them tuned and figure out who's going to have the high line, who's going to have the middle line, the low line, the bass line, and what the rhythms are going to be. And then what their textures are going to be and what are the parts that I'm going to interact with and what are the parts that are going to interact with themselves and trigger each other.

(Smith, 2016)

Producer Caterina Barbieri also uses modular synthesizers as the core of her musical system, interacting with pattern sequences and generative random operations from her modules to make music. "Playing with these machines," she explains

means actively interacting with this stream of sound, tuning yourself to an ongoing sound field and making it selective [. . .] What really interests me is finding an organic balance in my working flow between predictability and un-predictability, structure and fluidity, logos and chaos.

(Wilson, 2018a)

Barbieri's view of the production process perfectly illustrates the feedback loop connecting composer and technologies, in a circuit:

I see the compositional process as a system of feedbacks between technology and composer, as if it implied the creation of a kind of integrated cognitive circuit in which the design of the machine modulates the design of the musical thought of that specific person who "plays" it.

(Merrich, n.d.)

Like Hertz's approach to letting processes and equipment do the work of imagination while he assesses and develops its results, the way Barbieri works with her musical system harnesses the dynamics of complexity's happy accidents at the edge of chaos, building upon them:

Every form of computation requires the formal definition of a set of data to produce a larger body of output. You work within a closed system but then you define a process, a generative grammar able to generate an open system of possibilities . . . turning that practice of computation from being just a formal technique—an automatic procedure—into a creative process.

(Wilson, 2018a)

These examples illustrate how in electronic music production the producer steers the music, yet the system also steers the producer back by providing ideas upon which to build, sounds to respond to, and random operations to pursue or deviate from. Ideally then, *your music producing is a dialogue with the musical system you have configured.* But for this to happen, your system needs to be sufficiently complex to have nonlinearities and feedback opportunities built into it.

Like unpredictable weather or traffic patterns, a sufficiently complex musical system is able, notes Barbieri, *to generate an open system of possibilities* in the form of unanticipated compelling sounds to which one can respond through trial and error experimentation. As Allen Strange encouraged us 50 years ago in his study of electronic music composition, "Never wonder 'should I do this?' Instead an attitude of 'I wonder what will happen *when* I do this?' may lead to an expansion or development of a unique technique" (Strange, 1971: 5). Or as producer Beatrice Dillon puts it, "So much of this is run on complete naivety. Cast the net, pull out loads of stuff and then just see where it takes you" (Ravens, 2019). In sum, think about the production process as a co-generated complex system that generates unpredictable events and creatively volatile situations for you to interpret and build upon through your reactive artistry. The producer builds music from the ground up, reacting to a complex musical system that s/he has set in motion, moving towards discovery.

Examples of musical systems

Over time, every electronic music producer discovers and develops a musical system that works for them. For reasons of practicality, taste, and happenstance, one's system often revolves around the capabilities of a select few pieces of gear, such as a drum machine or synthesizer, and the constraints of a limited set of sounds. *Less is more.* To illustrate, we turn to a series of examples.

For his dub techno recordings *Pole 1-3*, the producer Stefan Betke (Pole) crafted minimalist tracks around the crackling and hissing sound of a malfunctioning Waldorf 4-Pole filter box. Released in 1996, the 4-Pole was originally designed "to process any audio signal with a 24db low-pass filter" as well as modulate the cutoff frequency and resonance of sounds (Waldorf, 2021). Betke describes how he came to use the hardware as the basis of a musical system:

> When I got this broken filter, it was connected to my mixing board, and at first I thought, *yeah, okay, it's not working, it does weird stuff, leave it there*. And then I forgot about it, and I was working on rhythmic parts with my [Roland TR-] 808 and other rhythm machines. And then, by accident, I unmuted the channel where this machine was connected to and I heard it and thought, this is much better than every rhythmical thing I programmed in my 808.
>
> (Grosse, 2020)

> It started making all these crackles and hisses—it was running in the back of a track I was working on, still with beats and everything, and I thought this is a nice atmosphere in the background, so I muted the beats and the crackles came more upfront and I thought this might be a good possibility to create something different—not working with beats but with randomly made crackles and noises.
>
> Since all the tracks were based on this broken filter that just crackles and hisses randomly this was very often the beginning of the recording of a track—that gave the rhythm structure, the main part. And then I simply started with some basslines or some atmospheres. In general, I'm really working more in atmospheres than in song structures.
>
> (Toland, 2009)

A second example is producer Andreas Tilliander (TM404). To make his music as TM404, Tilliander used a collection of renowned 1980s Roland hardware, including the MC-202 sequencer, TB-303 bass sequencer, and the TR-606, 707, and 808 drum machines. For him, this collection of Roland instruments "turned out to be a concept. Having all these machines—or rather instruments—playing at the same time" (Attack, 2013). Tilliander explains the merits of creating a unique system for one's music producing:

> When you can buy recorded hooks, drum beats and basses it means anyone can sound like you. If you put together a unique set of musical instruments and effects, you're bound to create something special.
>
> (Attack, 2013)

> The sound is everything. And by building a studio with various different echoes, multi FX, shitty digital synths, fantastic analog ones and so on, you will sound like nothing else.
>
> (Blanning, 2016)

In terms of both its equipment (e.g. software and/or hardware) and its sound palette (e.g. a single sound or many different sounds), a producer's musical system can be minimal or maximal. The producer Calum MacRae (Lanark Artefax) favors the kind of minimal system advocated for in this book: "I have a really basic set up. Just a laptop, software and a MIDI keyboard, they're all more or less essential as one another" (Inverted Audio, 2016). Producer and multimedia artist Mark Fell assembles a unique and minimal musical system for each project, the better to constrain his attention so that he can focus on how the system behaves and its components interact:

> My methodology is to focus on one or two specific bits of equipment and to explore them in detail. For any given project I usually assemble a collection of equipment and processes, and explore how they behave and interact. Rather than a universally perfect and utterly flexible studio setup, I opt for weird and peculiar.
>
> (Fell, 2016)

> I tend to be quite limited in the processes or technologies that I will use during the production of a record. So I won't think "I can add this and I can add that and maybe I can put this over the top." For me it's always about one or two elements that you keep re-explaining in detail, rather than adding another layer on top. It's about I look at the process and the technology and trying to just work with that, not confusing things by bringing too many ingredients to the production.
>
> (Wilson, 2018b)

At the maximal end of the continuum, Richard D. James (Aphex Twin), one of the most influential electronic music producers of the past 30 years, sets up a new musical system—which he calls "studios" and "setups"—for each track or collection of tracks:

> I realized I actually like making studios more than making music, because I like the possibilities of what you can do. I make these setups that will achieve some sort of purpose, so the way I've wired it together becomes the track itself.[3]

> If it takes you three years to set up a studio, and you've made one track with that setup, then the logical thing to do is not change anything and just do another one using the same set of sounds. Which I've done, and it's always really good because it's all ready to go. But I just can't keep it the same. I've always got to change something. All the tracks I've done in the last five years were made in like six different studios. It gets a bit complicated.
>
> (Sherburne, 2014)

One of the most fascinating and it-gets-a-bit-complicated case studies for thinking about musical systems is Autechre, the duo of Sean Booth and Rob Brown whose music sets a high bar for creativity in electronic music production. Since the 1990s, Autechre have programmed their own patches in Max, using these patches as an algorithm-based complex system to generate sound worlds that they develop into tracks. The group's recordings, notes Brown, are "just basically a modern output of our current rebuilt system . . . and some time spent with it" (Pareles, 2020). In numerous interviews over 15 years, Booth explains how their musical systems function, why they build and refine them, and how they interact with them to create music:

> We've built up a pretty extensive deep system now. . . . It's not like we make music, then use the system to replay it in new ways. The system itself is making the music each time, it's all about the capabilities of the system dictating what the music's like.
>
> (Muggs, 2016)

> There's no event generation taking place other than within the system we've designed. . . . How we play the system dictates how the system responds.
>
> (Tingen, 2004)

> We build systems because it's the most instinctive way of connecting things we invent. These systems do not produce a specific genre of music, they host modules, and manage all the protocols that allow them to communicate with each other. If we want to change the protocol, we have to enlarge the system.
>
> (Tsugi, 2018)

> We know what we do, and we know our habits. We built habits into the system that allowed us to be able to elaborate or expand the parameters or refine them, or perhaps closer define what we were after incrementally, and it made a lot of recordings happen.
>
> (Frame, 2013)

> It has always been important to us to be able to reduce something that happened manually into something that is contained in an algorithm. Then the algorithm allows us to add a bit more flair or a bit more deviation that we would also do ourselves in a little script. Just a few slight tweaks can spin it out into all sorts of recreations. It's a great way to spawn yourself if you like . . . and spawn your actions.
>
> (Pequeno, 2010)

In sum, for Autechre, musical systems, technology, and creativity are inextricably intertwined. Booth sums up this intertwining's crux perfectly: "it gets a bit hazy in terms of what's a musical idea and what's a piece of technology" (Sherburne, 2018).

In between these extremes of devotion to a musical system in the form of a single piece of gear or sound profile such as Pole or TM404, and complex generative music patches such as Autechre's system, are most electronic music producers who configure a system to help them make the music they want to make. Even if one does not have the resources to build new studios for each musical project, or Max programming expertise, this book recommends the DAW as an affordable, powerful, and protean musical system capable of creating enchanting sounds and the conditions for creativity.

Case study:

Brian Eno's musical systems of sonic treatments and generative loop machines

By the time Brian Eno was explaining the studio as a compositional tool, he had already been working with musical systems of one sort or another. In fact, the musical system concept is neatly illustrated by the sonic *treatments* Eno began implementing in his ambient music production work in the 1970s. Eno created these treatments by routing audio through a variety of signal processing and effects units that were connected in a circuit. These treatments were essentially bespoke musical systems with which to treat already recorded sounds and record the results. Software designer Sean Costello (founder of the reverb and delay VST plug-in company, Valhalla) explains how Eno set up "feedback paths through the mixing boards that incorporated several hardware processors" to shape sounds (Costello, 2011). To illustrate such feedback paths, Costello cites Eno's "shimmer" treatment patch. This patch was constructed by routing a sound output from a Lexicon 224 reverb unit through an AMS DMX 15-80s reverb with pitch shifting, then into a Lexicon Prime Time delay unit, an EQ on the mixing console, and finally, back into the 224. This circuit, notes Costello, "was just one of the 'treatment' topologies—[Eno] obviously put a fair amount of time into setting up different feedback configurations."[4]

Another facet of Eno's interest in musical systems is his work on generative music. In a 1996 talk on the topic, he traced its roots to the minimalist and process-oriented music of Terry Riley and Steve Reich. Riley's "In C" is a 1964 composition consisting of 52 bars of music, each of which are repeated any number of times by any number of musicians who proceed through the piece at their own pace. Eno noticed that the effect of the music's layered pattern juxtapositions created "a very complicated work of quite unpredictable combinations" (Eno, 1996). Eno also cited the influence of Reich's "It's Gonna Rain", a 1965 piece created using two copies of a tape loop of a preacher intoning "it's gonna rain" played simultaneously on two tape machines. As the machines gradually go out of sync, the voice transforms into echoing canons with strange textural-timbral qualities. Eno was inspired by both the music's sound and its process, noticing how Reich's use of tape machines was *a kind of system that transformed a simple input into a complex output*. The tape machines' going out of sync generated "a huge amount of material and experience from a very, very simple starting point." Eno heard in "It's Gonna Rain" an audio analog of a *moire pattern*, a visual interference pattern produced when slightly offset templates are overlayed. "It interested me that an artwork could be a system of amplifying detail," he noticed (Diliberto, 1988).

Building upon the music of Riley and Reich, Eno developed his ambient music concept, based on "a system or a set of rules which once set in motion will create music for you" (Eno, 1996). For his 1975 recording, *Discreet Music*, he configured two reel-to-reel tape machines into a "long delay echo system" through which to process sounds (Eno, 1975). In the liner notes, Eno described the album as "a technological approach to the problem" of how to create a musical system "that, once set into operation, could create music with little or no intervention" on the part of its producer. The producer's own contribution was limited to composing "two simple and mutually compatible melodic lines of different duration" and "occasionally altering the timbre of the synthesizer's output by means of a graphic equalizer." On the album's back cover, Eno provided "an operational diagram" of the tape machine-based system he used to make the music. This diagram is a kind of score that explains the processes by which the album's sounds evolve (Eno, 1975). In a way, the star of *Discreet Music* is the musical system used to generate it.

A final example of Eno's approach to letting sounds evolve on their own is *Music for Airports* from 1978, four compositions created using tape loops, some of which include three sung notes that repeat in cycles of varying lengths (roughly every 23, 25, and 29 seconds). In *Airports* and other subsequent pieces, Eno configured loops that never repeat the same way twice, instead generating ever new counterpoint and harmonies. Decades later, Eno's longstanding interest in generative musical systems culminated in the ambient music apps Bloom, trope, and Scape. From his sonic treatments and delay echo systems to his generative music experiments, Eno was ahead of the production curve in recognizing the creative power of musical systems.

Musical system lessons

Routing + simple rules = complex results, multi-centered action

One does not have to make explicitly generative or ambient music to implement Eno's ideas into electronic music production practice. First, the key idea is "how the system works and most important of all what you feed into the system" (Eno, 1996). In the context of a DAW, a producer might experiment with various musical systems via chains of connected effects through which to process audio or MIDI signals inserted onto tracks, effect Return channels, or even a project file's master output. This recalls Mr. Bill's *setting up a giant chain of effects that acts like a Rube Goldberg machine.* Such an effects chain can transform what is fed into it in unpredictable ways, the complex results of which can be built upon, through resampling, for example (see Chapter 6). A broader lesson from Eno's treatment techniques for enhancing sounds is that in the DAW, every sound can go on a journey somewhere. Eno's treatments are a forerunner of the many ways electronic music producers today route sounds through their DAWs. *A route is the way taken from a signal path's starting point to its destination.*

Routing sounds is one of the producer's most powerful techniques because it has no end. For example, a sound can be sent through a signal or effect processing plug-in, and the results of this routing then sent somewhere else. Or MIDI events from one track can be copied onto another track to control another set of sounds. In an Ableton video tutorial for Point Blank Music School, producer Freddy Frogs explains the origins of routing sounds in the analog studio and advises producers to experiment with their own routings in the DAW:

> In an analog studio we used to plug channels into other channels . . . and then worry about feedbacking the channel into itself. Experimenting with signals in an analog studio was essential. That is something that has been a little lost with software music. So try to use this routing matrix—the I/O section in Ableton Live—as a creative tool. . . . Experiment with the platform [the DAW itself]. Try to route things in different ways.
>
> (Frogs, 2014)

When we route sounds, it generates further ideas for how to route sounds, and as we listen to the effects of our routing we imagine yet more circuit paths to try. Within the DAW's ecosystem, one can also route sounds through and within VST plug-ins. Most VST synthesizers include a modulation "Matrix" page to organize such routing options that connects a modulation *source* (e.g. an LFO control) to one or more modulation *targets* (e.g. a Frequency Cutoff control). For example, U-he's Hive synthesizer has 12 slots in its modulation Matrix (Figure 2.2) and over 100 possible modulation targets (Figure 2.3). Such myriad possibilities for routing sounds from various sources to various targets create conditions for complexity through feedback loops that can lead the producer to unpredictable, cascading results. With the right treatments and routing, remarkable timbres become possible in the DAW.

A second lesson from Eno's work is to explore how "very, very simple rules, clustering together, can produce very complex and actually rather beautiful results" (Eno, 1996). Generative music,

Figure 2.2 Modulation Matrix for U-he's Hive VST synthesizer

Figure 2.3 Drop-down menu for U-he Hive's Modulation Matrix targets

he says, "specifies a set of rules and then lets them make the thing" (Eno, 1996). In a DAW we can implement the generativity idea via *automation*. For example, automating a parameter such as the Dry/Wet control on a reverb effect so that it increases or decreases at a certain rate over the time of a track specifies a changing set of transformations for the sound. Automating multiple parameters simultaneously creates compelling and generative-like composite sounds from simple materials. This recalls producer Boris Brejcha's point in Chapter 1, that through automation a track "starts to live, because the sound is moving a lot" (Brejcha, 2018).

A third lesson from Eno: generative music seeks to avoid "a single chain of command" and is instead "multi-centered" with "many, many, many web-like modes which become more or less active" (Eno, 1996). As producers, we can consider how to incorporate various levels of independent musical action into our tracks. Combining effects chains with automation of the parameters *within the effects in the chains* is one way to bloom multi-centered sonic complexities in a track. The modular Rack system in Ableton Live helps us visualize such sonic complexities. A Rack contains a chain of serially connected instruments and/or effects plug-ins with which "to build complex signal processors, dynamic performance instruments, stacked synthesizers and more" (Ableton, 2018: 283). One can even incorporate Racks *within* Racks to generate magnitudes of multi-centered sonic complexity. In sum, no matter what style of music s/he makes, a producer can use his/her musical system to generate layers of simultaneous, hard-to-predict activity via automated variations in a track's timbral, rhythmic, melodic, and harmonic activity.

An ideal, AI-based musical system

To create a broader perspective on musical systems, consider how your ideal system might work if it were AI (Artificial Intelligence)-based and could produce music based on your previous work and current state, drawing on all of the tracks you have made so far, plus other sounds you might be interested in. Such a system would have an *emergent* quality, meaning it would have "some ability to recognize the pattern that you gave it and complete the story, give another example" (Metz, 2020). On his music production podcast, Mr. Bill imagines such AI-based software helping producers to refine their musical system by observing their decision-making as they build tracks:

> I think you could have a program that just sits in the background of your computer just look-ing at the decisions that you're making all the time and just build an average mean for what it is you choose to do in every situation along the way. And after you've written a couple hundred songs and it has seen you do all of that, this thing could just be like, "I know your process" and just spit out new tunes.
>
> (Day, 2020)

While such AI-based software does not yet exist, thinking about how it might work illustrates the power of musical systems to generate complexity and help us produce.[5] Pondering such a system is a step towards learning how to make those sounds ourselves and, ultimately, curating our own creativity by reflecting on the decisions we usually make in the midst of our workflows. Ask yourself: *What would my ideal musical system consist of? How would it work? And how might I refine my existing system to bring me closer to this ideal?*

Setting up a music production system

Preferred ways of working and defining a musical system

When you are setting up your musical system, consider your preferred way of making sounds. You may already be a keyboard player, a drummer interested in applying your skills to finger drumming on a pad controller, a DJ interested in remixing, or a newcomer to production. Even if you do not play a musical instrument, that is no hindrance to producing. In fact, many pro-ducers who do not play instruments work with samples, field recordings, loops, and sequences. What is important is having a way to get information into your musical system, to load the sys-tem with input so that you can set feedback loops in motion and engage with the possible com-plexities or output that results. Try using a MIDI keyboard for playing melodies and chords, and finger drumming beats. If the controller has rotary knobs or faders on it (Figure 2.4), these can be *MIDI-mapped* to the DAW software to control parameters within it. Producer Robert Henke (Monolake) defines musical systems in terms of a set of tools that interact. Henke uses "tools" broadly to mean any combination of software or hardware instruments, libraries of sounds, and ways of treating and routing these sounds. "It has indeed become a requirement for every electronic musician: define your system," he says. "Essentially it's a quest for identifying those five tools that interact best with each other for what you want to achieve" (Goldmann, 2015: 46).

Figure 2.4 Arturia 61-key MIDI controller

Keyboard and Grid pad controllers

Many electronic music producers work with some kind of controller to trigger samples and input MIDI into their DAW. At the time of this writing, the most popular devices are MIDI keyboards and Grid-style drum pad controllers.

The generic MIDI keyboard controller (Figure 2.4) is often referred to as a "dummy" controller because it contains no sounds of its own and must be connected to a computer.[6] In its standard 25-, 49-, 61-, and 88-key configurations, the keyboard is merely an input device for triggering whatever sound you have loaded into your software. Use the dummy-ness of the MIDI keyboard controller to your advantage! Unlike hardware instruments, the controller's sound is not defined until you assign a sound to it in the DAW.

Despite being without sounds, the layout of the MIDI controller is that of the piano, with white and black keys and 12 semitones per octave that tacitly connect the producer to the long tonal history of European music. In this way, the keyboard materializes musical possibilities in the form of potential pathways for the fingers. If you are not a keyboardist, you can still use the instrument's notes being close at hand to your advantage. Try out different chord shapes: if you can play a C major triad (notes C, E, and G), you can also transpose this hand shape elsewhere on the keyboard to make different chords (such as an e minor or a b minor diminished triad). *Experiment.*

The most influential controllers of the past 20 years have been Grid Controllers with rows of touch pads arranged in 4-by-4 or 8-by-8 grid matrix formations. The Grid has its origins in Roger Linn and Akai's MPC60, a sampler developed in 1988 which we explore more in Chapter 5. Examples of Grids include the MPC, the Novation Launchpad, the Monome, and Ableton's Push. Since Grids are not keyboards with notes laid out low to high, left to right, their design frees the producer to approach melodies or chords in novel ways. Grids are also ideal tools for step sequencing beats: their touch sensitive rubber pads are optimized for finger drumming and their layout displays the beat subdivisions and flow of musical time as a sequence of flashing lights.

Templates and sound sets

A template is your default set up in your DAW that includes, at minimum: pre-configured audio and VST instrument MIDI tracks, effects loaded onto Return tracks, and your go-to signal processing chains or Effects Racks at the ready. One reason to use a template is to have the essential elements of your musical system present whenever you work; "a well thought out project template will enable you to get up and running in a single click" (Walden, 2010). Some DAWs come pre-loaded with production templates. Steinberg's Cubase, for example, includes "Dance Production," "Hip Hop Production," and "Metal Production" templates. But make your own. As a start, explore your DAW and VST plug-ins to find preset sounds that you like. Modify these sounds, save them, and make them easily accessible when you are working.

Equipment doesn't matter (that much)

From the latest software instruments to re-issues of classic analog synthesizers, there are a plethora of tools with which to produce electronic music. Sounds are cheap and easy to come by, computer memory is abundant, and there are more instrument and effect plug-ins than a producer will ever have enough time or money to try. One is forgiven for thinking that production is about its gear, because there is so much extraordinary gear to use. But consider that *the specifics of your equipment are relatively unimportant*. As Mark Fell reminds us, "don't worry about what technology you need. Use what you have" (Fell, 2016). It doesn't matter that much what equipment you use because as a producer your primary job is to *develop ways of interacting with your musical system that inspire you and generate compelling sounds*. The late keyboardist and composer Harold Budd, one of ambient music's pioneers (and a quiet inspiration for this book), speaks to using this approach to recording. Budd used whatever synthesizer was *closest at hand*: "I have a philosophy that one is obliged to use what's there. You don't need an awful lot of stuff, but you use fully everything that's there" (Goldstein, 1986).

Conclusion: Approach and Quest

This chapter has explored the importance of setting up complex musical systems and workflows for interacting with them. This brings us to Approach and Quest. Approach is the way you draw on your tacit knowledge to interact with your musical system. To illustrate the concept, consider the deceptive fact that many producers have a favored piece of gear because of its sound, its workflow quirks, or because of how supposedly intuitive it is to use. A producer may swear by a particular technology, such as a Korg MS-20 for basses, or an Elektron Octarack for sequences. But remember that a producer's enthusiasm for gear glosses over the fact that s/he has learned meaningful ways of using and interacting with it, *having discovered through trial and error uses that makes it feel indispensable*. As producer Darren Cunningham (Actress) puts it, "once you've decided on the equipment you're going to use, you have to learn it, understand it" (Duplan, 2017). Such is the way electronic music production is always a dialog between a musical system and a producer's tacit knowledge of how to make use of it. In sum, your Quest is to get to know and develop both the potentials of your system and your skills for using it to create the music you want to hear. If you are unsure what to use, consider the DAW as your primary production tool. Software is affordable relative to its power, uncannily models the capabilities and sounds of hardware, and is elegantly compact, compressing a studio into a laptop.

Production Quest

1   Start simple. In your DAW, set up a musical system comprised of three tracks of synthesizers, each for one type of sound: chords, bass frequencies, and percussion. Make a piece (or ten) with this system.
2   Expand this three track-based musical system by adding three Return channels for effects. Insert a different effect on each Return channel so that the effect can be used in various amounts on each of your track's three elements. For example, Return 1 can be a reverb, Return 2 a distortion, and Return 3 a delay. Shape the music you made in Step 1 by sending

your effects Returns in varying amounts to each of the sounds, automating the effects so that they do different things over time. Explore the parameters on the effects as well, automating them to bloom Eno's "multi-centered" complexities into your track. Explore the creative possibilities of this simple yet complex musical system with three sound types and three effects Return channels. Save the system as a DAW template, and build a new one tomorrow.

Notes

1 Research on the musicology of music production has examined these paths. For example, Aksel Tjora (Tjora, 2009) applies concepts from social construction of technology studies (Bijker et al., 1987; Akrich, 1992) to frame musicians' creative workflows in terms of the "usage trajectories" they construct within the implicit "scripts" of their electronic musical equipment (Akrich, 1992). Examining uses of the Roland MC-303 Groovebox, a hardware sequencer and drum machine popularized in 1980s electronic dance music, Tjora shows how the instrument's scripts accommodated musicians going off-script by "constructing their personal user trajectories" (Tjora, 2009: 175). Similarly, in their history of the Moog synthesizer, Trevor Pinch and Frank Trocco show how scripts "try to contain the agency of users, but users can exert agency too by devising their own alternative scripts" (Pinch & Trocco, 2004: 311). Indeed, there is often a divergence between what manufacturers intend for their equipment – their implicit scripts – and how musicians deploy it. For example, with regard to drum machines such as Roger Linn's LM-1 and Roland's TR-808 that changed the sound of music in the 1980s, their user trajectories were largely uncharted, and there was no script to guide musicians in how to use them. Musicians devised workflows for using these new instruments arrayed into musical systems and through that discovered new sounds.

2 This idea is echoed by mathematician Marcus du Sautoy in his book, *The Creativity Code*: "Being creative requires a jolt to take us out of the well-carved paths we retrace each day. This is where a machine might come in: perhaps it could give us that jolt, throw up a new suggestion, stop us from simply repeating the same algorithm each day. Machines might ultimately help us, as humans, behave less like machines" (Du Sautoy, 2020: 4–5).

3 To illustrate: the studios and setups James configured to make his 2014 recording *Syro* used some 138 individual pieces of hardware and software equipment (Minsker, 2014).

4 (Costello, 2011). Perhaps taking inspiration from Eno's treatments, one of Costello's Valhalla reverb plug-ins is called Shimmer.

5 For an overview of uses of AI in music production, see Drake (2018).

6 In *Sonic Writing*, Thor Magnusson asks us to imagine what it would feel like *being* such a digital instrument: "Unlike the acoustic instrument that knows its user very well, your user behaves differently every time you are played, and that is possibly because you are never the same either!" (Magnusson, 2019: 35)

3 The faintest traces

Beginnings, approaches, improvising, and being lost

Introduction: begin anywhere

To begin a music production project is to face the fullness of possibilities and the emptiness of what could be done but has not yet been tried. The composer John Cage once famously advised, *begin anywhere*. He consulted the *I-Ching* and rolled dice to generate his compositional moves and help him decide what to do, finding a process (Cage, 2015: 32). He added, *there's a temptation to do nothing simply because there's so much to do that one doesn't know where to begin* (ibid.: 32). For Cage, the purpose of artistic practice was not to express the self but to quiet the mind. We might follow the constraints of outside energies, he said, to let our creative process be more like nature *in her manner of operation, complete mystery* (ibid.: 25).

The *begin anywhere* idea is advice for the electronic music producer to get going, make sounds, and explore the accidental, unforeseen experiences that materialize from the process. My process involves a computer and DAW software. There is a black box magic to producing electronic music this way, in that I co-create with tools whose workings always extend beyond what I know about them. The computer and its software run computational processes behind the scenes. Their fast processing is almost magical, creating "a wealth of sound seeming to emerge from a thin wedge of electronics" (Bach, 2003: 3). Working with my system, I get busy with making sounds.

This chapter outlines the experience of beginning an electronic music production project and offers some general concepts that producers might consider as they embark on their own work. I explore the connection between our musical system and our ideas, the value of focusing on single elements in the system, compositional placeholders and through-lines, and creating variations through permutations of limited sound sets. Then I turn to improvisation as a way of negotiating a musical system through productively getting lost and un-lost. I conclude by considering the impact of mindset, the value of creative outsourcing, and steps for resisting the musically predicable.

My *begin anywhere* is recording marimba, playing sequences of chord rolls in a minor key (Figure 3.1). To make the *begin anywhere* a somewhere, I had decided upon a few constraints ahead of time:

- a tempo of 112 BPM
- 16th-note subdivisions for the chord rolls
- each chord rolls for 8 beats
- fade out the rolls over each 8-beat span

I improvise chords in all registers of the marimba, from high to low. Some chords have open voicing with large intervals between the notes, and some have closed voicing with the notes clustered together. Somewhere in these improvisations I hope to find material upon which to build.

Figure 3.1 The author playing marimba

Maybe I could arrange the chords into new progressions, layer them, or cut them into smaller samples? I can figure this out later.

I am always curious about how other electronic music producers begin making a track, because the possibilities at the outset seem endless. As the producer Beatrice Dillon says, "It's crazy that you can do anything, and that's what gives you a headache" (Turner, 2020a). And does how one begins even matter? Does it determine the direction the music will take, or simply get the proverbial production ball rolling? Producer Geir Jenssen (Biosphere), known for his ambient electronic music and field recordings, begins with a quality sound, either sampled or synthesized, found or made. For Jenssen, a quality sound can contain within itself an entire composition:

> I often start with a sample taken from a record or a film. Sometimes I start with one of my own field recordings, or by making new sounds on a synthesizer. . . . A good sound can often be more important than the actual composition. The better the sound, the less you have to "compose." A good, organic sound can be a composition in itself.
>
> (Fischer, 2012)

Sometimes Jenssen begins by putting a sample into software that will mangle it, revealing hidden sounds within. He has used Argeïphontes Lyre, quirky software created by the electronic musician Akira Rabelais, that generates unpredictable sounds out of audio loaded into it.[1] Jenssen describes being surprised by the results:

> I can, for example, upload a track with some old Ukrainian women singing. What comes out is something completely different—there was one track that sounded like Elizabeth Fraser from Cocteau Twins. You always get these surprises; you never know what comes out. So

I just picked the best parts of it and re-sampled that, and made the tracks from these short snippets.

(Ryce, 2016)

Similar to Jenssen's focus on a quality sound, the producer Benjamin Wynn (Deru) searches for "the smallest idea" with which to begin:

The hardest part of the writing process for me is starting, so anything that I can do to get started is great. The smallest idea can get me going in terms of engaging in the process of work, like putting some samples in a granular synth . . . or messing around with a new plugin or melody.

(Raihani, 2017)

The producer Ryan Lee West (Rival Consoles) sometimes begins by sketching ideas for tracks on paper, away from the DAW (Figure 3.2). On his website, West shares photos of track idea sketches along with an explanation of his process:

Over the past 5 years I have been sketching compositions on paper. Sometimes they are detailed, specific outlines of what I imagine for the music, sometimes they are how I would like a synth to sound, sometimes they are me thinking out loud about the structure of the music I am working on. The main reason why I do this is because it is a way for me to problem solve away from the computer. I find the computer is so powerful at trying things quickly that it can get in the way and overpower some of my decisions.

(West, 2020)

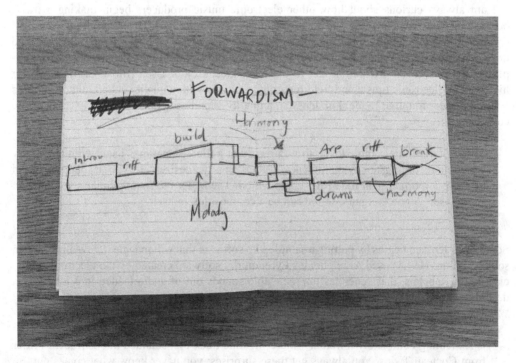

Figure 3.2 Rival Console's "Forwardism" track idea sketch (Drawing by Ryan Lee West, rivalconsoles.net)

Finally, the producer Darren Cunningham (Actress) re-frames the production problem from how to begin *making a track* to simply *making sounds*:

> I don't really sit down and go, "Alright, I'm gonna make a track today." I just go, "Alright, I'm just gonna make things, and I'm gonna use different things to influence what that sound does."

> (Darville, 2017)

I drag the audio files of my marimba playing into Live and load Simpler, one of Ableton's samplers (Figure 3.3). I assign each marimba chord sample to a pad on my Grid controller and trim the start and end points of the samples, so that when I tap a pad, a marimba chord plays. As I assign the chords to the pads, the pads light up, as if suddenly alive and radiating my own excitement back to me. What were once different mallet-holding hand positions over an acoustic instrument are now disembodied chords on the Grid, awaiting my next decision. I can re-play the marimba chords by tapping the pads, free of the intentions of my original performance. On the Grid, the chords need not remain in the order in which I had played them. The producer Sam Shepherd (Floating Points), describes a workflow whereby he plays chords on a keyboard, samples them, and then re-sequences the samples:

> I record lots and lots of chords completely independently of each other. The overall key doesn't matter, they can have no relation to each other whatsoever. Then you sample these chords, lay each one out on a key of a keyboard and experiment with playing them back in a different order to see if a certain combination sounds nice to you. . . . This technique can get you exploring harmonic progressions that you wouldn't have thought up.

> (Smith, 2019)

Now the Grid is an instrument for re-playing the marimba sounds: I *tap-tap-tap* the pads in different sequences and rhythms, finding a flow, learning how the chords on each pad relate to one another (Figure 3.4). At this early stage of the production process, the pleasure is uncertainty and the task figuring out the relationships between the pad sounds. There are so many permutations for playing the pads together or sequentially that I need to whittle down their possibilities to just a few. I try out various chord sequences and follow those that sound compelling. *What's that sound? Play it again!*

With the uncertainty of not knowing where each marimba chord is on the Grid, I tap to hear the sounds, each pad tap triggering a two-measure long series of 16th-note pulsations. "Playing" the marimba by tapping pads is an uncanny sensation because all of the embodied physicality

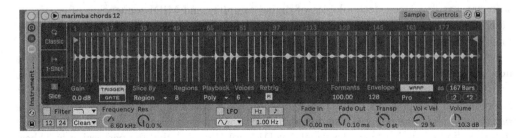

Figure 3.3 Marimba chords loaded into Ableton Live's Simpler

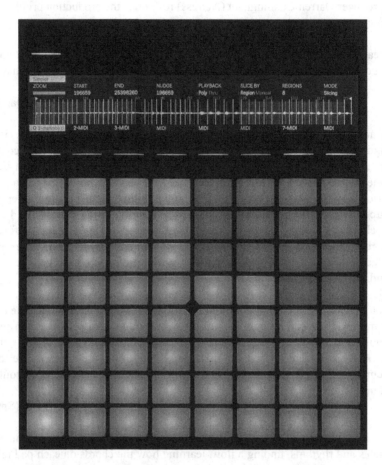

Figure 3.4 Marimba chords loaded onto Ableton Push pads

of playing the acoustic instrument—hands gripping mallets, mallets striking wooden bars and feeling their give, arm gestures changing to alter dynamics and tone—has been compressed into a free floating moment I can trigger however I wish. The Grid has increased my musical options by magnitudes: if I move quickly between two pad taps, the marimba chords overlap like waves in a way that was impossible to achieve when I recorded them (and would require two marimba players playing together). As Alan Durant presciently noted in 1990, a decade before my DAW software was developed, technologies of digital sampling and MIDI sequencing offer, among other things, "solutions to problems of inaccurate, incomplete, or technically impossible performance" (Durant, 1990: 185).

My pad tapping and listening for relationships among the pads soon becomes improvising a new sequence of marimba chords. I tap to hear which pads hold the most unusual chords, all the while trying to remember which chords are where and wondering how they could go together. A production task presents itself: *How can I devise an interesting chord sequence using the pads I have under my fingers?* There is nothing new about my process; I am, literally, tapping into sampling's long heritage. We remember DJ Kool Herc isolating the percussion breakbeat sections of records and hearing a new groove, or Pierre Schaeffer stitching together recordings of pulsating train sounds to make his *musique concrète* in which "the train must be forgotten and only sequences of sound color, changes of time, and the secret life of percussion instruments are heard" (Schaeffer, 2012: 14). As I tap the pads, I hear what I could not hear when I recorded the marimba. Now the instrument sounds unfamiliar. Its sounds are a flood of information, a sensory overload.

If I could, I would somehow try out every combination of pad tap patterns, but I can only work with what I happen upon in the moment, committing to it, and following where it leads me. I find a chord that sounds like a beginning, then listen to and compare how chords on the other pads sound following it. From one chord to two, then three and four, I build up a sequence and record it, tapping on the pads along to a metronome click (set to 82 BPM, the tempo at which I recorded the marimba). I leave space after each tap to let the 8-beat marimba pulsations fade out. I like odd-numbered, long sequences—not 4 or 8 bars, but maybe 23 or 42—because their complexities keep me surprised.

Now that I have begun, each day I repeat the process of assigning marimba chords to the pads on the controller, improvising a long sequence by tapping the pads, and recording the results.

With my sequences on the screen, I move them around to hear how variations sound. I drag the sequences further apart to create more space between them and elongate the progression, I play with the order of the chords within a sequence, or I duplicate one chord and delete another. Each of these simple moves can create unfamiliarity and open-endedness, which I hope to preserve as long as possible. "I start with ideas," says the producer Stenny, "but often they morph into something that I still don't know how to define, and I don't intend to do so" (O'Gara, 2019). Re-ordering the sequences creates sonic surprises that I could never have stumbled upon any other way, and listening to these variations I realize that there are always other angles from which to approach one's material. "I just like to get the idea down," says the producer Imogen Heap, "and play with it once I get it in the computer" (Halle, 2010).

The sequences are my *begin anywhere*, a foundation to build on and play with by trying out other parts and other sounds in pursuit of some kind of energy. But how does the producer know what sounds go with what?

Musical gestures, musical systems, and discovering dialectics

In electronic music production, there are endless ways to begin. I began with marimba chords, but I could have begun by sampling a single marimba note and worked with that—using its waveform to make a tuned drum kit or granular pad, for instance. Sometimes we pursue a production gesture because it makes sense of the situation in which we find ourselves: we improvise chord rolls, played at 112 BPM, fading out over eight beats. A production process is a commitment that connects us in a tactile way to where we are *now*, and, as long as we pay attention, it will offer us options for moving forward.

The composer Arvo Pärt describes composing as identifying the "single gesture" that is inherent in a piece of music that has yet to be created and then finding a system through which to express it:

> A composition comes as a single gesture which is already, in essence, music. . . . The compositional task is to find the appropriate system for the gesture.
>
> (Hillier, 1997: 201)

This idea of connecting musical systems to musical gestures is germane to electronic music production, and helps frame our work. The producer Ben Lukas Boysen explains Pärt's influence on his own productions, specifically Pärt's belief that every sound requires good reason to warrant inclusion in a composition:

> Arvo Pärt—who's one of the few composers I admire endlessly . . . said that he's trying to "decorate" his music as little as possible—everything in there should have a meaning, function and reason to be there—this impressed me deeply and I fell in love with the idea of music as "acupuncture", where every element hits a certain nerve and where there are no fillers, no needless elements.
>
> (Fischer, 2014)

In electronic music production, a production gesture can be many things. It can be something musical—a melody, an arpeggio, a bass line, a chord sequence, a beat or rhythmic texture, or an effects chain. Or a gesture can be an extramusical sensation like a mood, a feeling. While Pärt does not explain what gesture in music is, imagine its impetus coming from anywhere. When it appears, its form is the DNA of what could eventually become a finished piece. Ryan Lee West describes something akin to Pärt's single gesture idea when he speaks of figuring out a track-in-progress's most essential aspects:

> Many pieces take me a long time to understand them, or at least understand what is the most essential aspect of them that then dictates the logical progression of the composition. So I need a lot of time just listening to my ideas.
>
> (West, 2018)

The second half of Pärt's idea—composing as finding the "appropriate system" for the musical gesture—is also useful for music producers to think through. As we saw in Chapter 2, a musical system is an assemblage of technologies (e.g. DAW software, VST plug-ins, guitar pedals, etc.) connected to work as a unit so that recorded, sampled, or synthesized sounds can be routed through them. But a musical system can also be thought of more expansively as *a concept in search of an action*. A system can be a general idea—for instance, an expression of musical density or emptiness—that the producer finds a way to manifest in sound. Or a system can extend beyond technologies and musical concepts to draw on the producer's life in, and outside of, producing music.

Consider, too, all of the possible *dialogues* in production—between the producer and a musical system, between the parts in our tracks, between our tacit knowledge and what we do not yet know, between our practice and our potentials. John Cage used another phrase, *discover dialectics,* to describe these conversations. The scope of any electronic producer's musical system is vast and its workings complex because its components are in continual dialogue with one another as our questions become musical answers, leading us to yet more questions as to what to do next. For the producer, discovering dialects is figuring out possible points of connection in a musical system, hearing the simultaneous conversations happening on different strata of experience and awareness—between sound and feeling, technologies and techniques for using them, influences and ideas for putting them into action, and responding to what the music seems to need.

Summing up so far, at the outset of beginning every new track a dilemma presents itself. While we can *begin anywhere*, we need to closely attend to our initial ideas—whether the product of deliberate decision-making or random *happy accidents*—and permit them to suggest a direction. At the same time, our tinkering with sounds, parts, and structures from the get-go reveals new possibilities for our material. Something is always happening, but are we noticing? And if we notice, how will we respond to proceed? Let us turn to the work of some other music producers for guidance.

Jon Hopkins' ways of proceeding

The music of producer Jon Hopkins has been described as "fusing technology to humanity" in that his tracks "evolve and grow like an organism, never getting stuck in flat, symmetrical loops that afflict some other intrepid button-pushers and laptop musicians" (Nguyen, 2018). In candid interviews about his workflow, Hopkins says that the initial stages of generating ideas for a recording project are difficult, because he understands how much time and effort are required to produce a finished track:

> The amount of layers and processing that end up in the final versions, and all the things that have gone into making it sound just so—there's really a huge amount of time and work. So when you've got nothing at all it's daunting. . . . Of course, all you have to do is start.
>
> (Smith, 2018a)

Placeholders

Hopkins' strategy is to work quickly at the beginning of a project to capture ideas before over-thinking them. For example, he sometimes begins by drumming a beat on his desk, recording and tinkering with it a bit. Then he plays chord *placeholders* along to it:

> For the sake of immediacy, I might drum a beat on the desk with a mic and split that out onto the Push. Then maybe I'll resequence and EQ it into the semblance of a beat, then quickly play chords over the top. . . . There's something to be said for working with placeholders so you can move on with the overall composition.
>
> (Smith, 2018b)

Hopkins improvises many parts and records his performances as audio and MIDI clips in Live's Session view. "I tend to catch on live performances really quickly" he says, "and improvisations get put down in the different clips" (Rancic, 2018).

Another angle or through-line

In Hopkins' experience, musical placeholders can sometimes be springboards for new production paths he calls *through-lines*:

> What's interesting is that sometimes you might find that some element of the placeholder, like its reverb for instance, can be the genesis of another angle or through-line of inspiration for how the sound progresses. There may be a tiny bit remaining of the very first sound. You get addicted to what you start out with. If you remove every trace of that original spark, sometimes the track can suffer.
>
> (Smith, 2018b)

Once an element in the music—such as a reverb—is identified as a through-line, the producer can tinker with it. Hopkins finds this pursuit *exponentially interesting*:

> There's always something that you can fiddle with, and that will trigger another idea, and before you know it you've resampled what you've just done, like the tail of a reverb

accidentally you'll just hear it and straight away resample it and then you've got a whole new sound to play with. It's sort of exponentially interesting, and hopefully that's coming out in the tracks.

(Heaton, 2016)

Harold Budd's three steps

After an earlier career as a drummer and composer of avant-garde classical music, in the 1970s Harold Budd turned his attention to composing ambient music and playing piano. Budd's artistic influences include John Cage's essay "Where Are We Going And What Are We Doing?" (Cage, 1973: 194–259) as well as the abstract expressionist art of Robert Motherwell, who espoused the practice of "automatism," automatic doodling to tap into ones unconscious.[2] Searching for his own creative principles, Budd developed a simple and minimalist approach to improvising piano and recording that applies to navigating electronic music production's ocean of workflow options. As a complement to Cage's *begin anywhere*, Budd used a three-step strategy for moving forward, no matter where he may have begun. First, he *focuses on a single and small musical thing*: "The way I work is that I focus entirely on a small thing and try to milk that for all it's worth, to find everything in it that makes musical sense" (Goldstein, 1986).

Although Budd does not say what this *small thing* is, we can imagine possibilities from our own experience—for example, a phrase, an interesting sound, a rhythm, an effect, or the unexpected juxtaposition of two marimba chords. The details of what the small thing is do not matter as much as the urgency that *we find something to focus on*. Budd's second and third steps are connected: to *extract from this point of focus all of its value, by uncovering everything in it that makes sense*. "I really like to find as much life as possible in the smallest amount of material," he explains. "A very simple scale, a relationship of note against note, especially a sustained note" (Goldstein, 1986).

To further illustrate Budd's approach, consider the plethora of preset sounds available to the electronic music producer, a topic explored more in Chapter 4. Despite our having an excess of sounds within our DAW, we can chose to begin with a *single* sound that catches our ear. We might ask ourselves: *What do I find enchanting in this sound, and what could I do with it to reveal its potentials?* A producer can apply Budd's focus-and-find-everything-in-it-that-makes-musical-sense approach to audio samples, MIDI sequences, drum loops, sound design, or ways of effects processing in the DAW.

Budd used his approach to make his recordings, keeping his process as streamlined as possible. In his collaborations with Brian Eno in the 1980s such as *Ambient Volume 2: The Plateaux Of Mirror*, released in 1980, Budd would find a synthesizer preset, make a few changes to it, and get playing. As we saw in Chapter 2, Eno routed the sound through an electronic *treatment*—for example, a chain of reverb effects—he had devised to create a specific timbral aura. The treatment shaped how Budd played by literally processing his improvising. "The sort of treatment you hear on the piano influences exactly the note-to-note process," Budd explains. "The length of time between musical gestures, and the kind of taking advantage of the ringing timbres which I'm very fond of" (Goldstein, 1986). Eno noticed how Budd interacted with the sonic complexities of his treatments, hearing how Budd' playing played with their effects:

I used to set up quite complicated treatments and then he would go out and play the piano. And you would hear him discovering, as he played, how to manipulate this treatment. How to make it ring and resonant. Which notes work particularly well on it. Which register of the piano.

(Diliberto, 2016)

In sum, once Budd found a keyboard sound he liked, he committed to the sound, and recorded with it as is. "My idea of working in the studio," he says, "is to play as very little as possible, and take what you got and then work with that" (Diliberto, 2009).

Play as very little as possible, and take what you got and then work with that. Now that I have marimba chord sequences, I begin trying out other parts and other sounds. In the DAW there are hundreds (thousands?) of sounds that could potentially sound good with the marimba chords, but on this project I constrain myself to just a few sounds that fit. A sound that fits is often a contrasting sound—a sound that complements the timbre of the marimbas, or that occupies a different frequency stratum. *Pay attention to contrast.*

Improvising to find something interesting

I'm playing along to the marimba sequences, improvising on the keyboard, playing one part after another, from a beginning to an end several minutes later. Improvising long parts gives me the best chance of capturing a performance with energy, finding a form, and generating material that can be shaped and edited later on, a topic we explore in Chapter 7. The keyboard is familiar terrain, and my hands find chord shapes, pathways, and places to go within its "field of keys" (Sudnow, 2001: 51). I improvise in search of a mood and some kind of interestingness that holds my attention, but since my hands fall into predictable patterns that have worked in the past, I aim for what I haven't heard before. I play to hear where the playing takes me, moving up and down chord inversions, trying out overlapping dissonances, waiting for a compelling sound. A few chords catch my attention, so I repeat them. Is this repetition enough to build on?

As I audition sounds, I am simultaneously improvising chords, melodies, and bass lines. Like the sequenced marimba chords, these new ideas are also placeholders—attempts at filling out the music in a preliminary way with contrasting timbres. Even so, I try to make temporary parts usable. As Hopkins points out, sometimes traces of a placeholder will survive into a track's final version. Every sound and part I try is potential scaffolding to hold up something else later on in the production process.

Improvising strategies: call and response

I improvise to come up with new parts, listening to what I already have and then trying to complement that by playing something against it. Improvising is a jam session with a clone of myself, and I find that jamming works best when you hear the musicians interacting through a call and response dynamic. Layering parts in this way helps the music sound lively and tensile—as if it is listening to itself on different layers of rhythmic and textural action. As I layer chords, melodies, and percussion that respond to one another, the track's texture becomes a more animated and oscillating experience in which every sound has its own pulsation.

Accidental counterpoint, finding variations

Over the course of sequencing the marimba chords, looking for contrasting sounds, recording placeholder parts, and improvising calls and responses, errors appear in my workflow. I drag a marimba sequence a few beats too far over and now one of its chords overlaps with another sequenced part in an odd way. But hold on—it sounds good! The overlap creates an interesting accidental counterpoint. I leave the overlap be, for now, as another placeholder. Such accidents can sound more interesting than the results of my deliberate decisions and when they do, I welcome them. At the same time, chance is not a complete substitute for intentionality: the surest way to make a track sound compelling is to shape it by deliberate adjustments.

As the electronic music producer sketches a rough form for the music, s/he questions its sound at each of its passing moments, listening to what s/he has so far and wondering about how to organize it into a more cohesive and intentional whole. The producer Jerrilynn Patton (Jlin), whose rhythm programming we explore more in Chapter 5, speaks of her process of finding permutations and combinations of a limited sound set to create variations:

> Basically, it starts with two sounds and it really becomes just a thing of permutation and combination. Because what happens is, it starts off with maybe three really simple variations and what ends up happening is these three simple variations become nine very complex sequences and ideas. All the variations are still there, just switched so many times it sounds totally different now.
>
> (Yoo, 2017)

I have begun and found some ways to work, yet what makes electronic music production complex is the number of possible interactions between what I notice in the sounds and the musical system at hand. At any moment, I could morph a sound to be *totally different* and my understanding of what the track could be will shift accordingly. The production process requires one to be intentional and systematic, yet also open to letting unintended things happen. Which brings us to being lost.

Being lost and getting un-lost

Throughout the production process I am often lost in the sense that it is not clear where the music is going and which steps I should take to develop it. At times, every moment of the experience seems ambiguous. For example, I notice myself half understanding how a software instrument works whilst trying to get an interesting sound out of it; playing a meandering chord progression and not knowing what comes next; mixing effects together in an effects chain and not being able to predict how it will sound; turning a chord into an arpeggiation set to "random" mode and not knowing which notes will hit and when; being uncertain as to whether the track needs three parts, 12, or 20.

Producers are perpetually trying to engage their momentary ignorance of how to understand a musical system and how to use its capabilities to unlock new sounds. Perhaps then, being lost is the essence of the electronic music production experience? If this is so, then our ongoing task is to figure out ways of bridging the gap between what our system is offering us in the moment and what we notice of that offering. The results may be uncertain, but just because we cannot predict an outcome is no reason not to explore. Being lost is a Houdini Opportunity, where the trick is figuring ways out of the locked boxes of our own making.

A solution is to *improvise*. Improvising is a way to explore being lost. It is the spark that ignites an idea for a track, the sudden left turn that steers us elsewhere, the *What if I do that? or How about this?* questioning that reframes a production dilemma, the resourceful making-do with whatever sound is at hand.

Electronic music producers incorporate improvisation into every aspect of the production process. By improvising with a musical system to build a track that is as alive as possible, the producer moves the music forward and in the process gets un-lost. Improvising can be applied to anything in a production project: you can improvise by playing with a track's sound design and timbres, its grooves and beats, its sequences and loops, its melodies and harmonies, its arrangement, and its mix. Improvising has no technical or theoretical prerequisites and might well be the producer's most widely useful skill, compressing composing into a fleeting instant, distilling decision-making into evanescent moments of commitment for capture.[3]

Improvising also turns up the volume on our tacit knowledge, bringing to our awareness what we didn't know we already knew by drawing on our embodied experience. As Deckle Edge notes in his study of craft, "the most versatile and the most complex piece of kit we have at our disposal is our own body" (Edge, 2019: 24). When you improvise, you surprise yourself with what your body knows. You go where you have never been before, get lost, and then figure out ways of getting un-lost by turning where you are into a place you want to be. Improvising is how a producer explores routes through a musical system, trying things out. Through improvising, the producer's thinking evolves. Above all, improvisation generates plenty of potentially useful material for a track.

Improvising the sound of a virtual band freeing themselves from musical conventions

The varieties of improvisation carried out by electronic music producers can be illustrated with an analogy to an acoustic band improvising. The power of this band's improvising is that anything could happen, and the musicians might pull off feats of virtuosity and tuned-in togetherness that leave you in awe as you wonder: *How did they do that?* This is the level of performance on classic live recordings that capture fleeting moments of enchanting and dexterous playing, thinking, and ensemble listening. Think of Keith Jarrett's sprawling and meditative *Köln Concert*, or Miles Davis's restrained and minimalist *Kind Of Blue*, for instance. The best performances, notes Barbara Gail Montero in her study of expertise, "allow observers to witness some deliberate, conscious thought in action" (Montero, 2016: 140).

In theory, an electronic music producer could approximate this real-time interpersonal interaction, fleeting magic, and conscious thought on the fly by layering parts recorded once and so perfectly that they would not require any further editing or manipulation. But producing does not require the real-time virtuosity traditionally ascribed to skilled instrumentalists. As the producer Guy Sigsworth observes, "we have great artists like Aphex Twin, who are, in a sense, non-real-time virtuosi" (Fischer, n.d.). The producer has many non-performance-based ways of creating sound worlds that feel alive. In fact, most electronic music tracks contain accumulated traces and layers of performance from their non-real time making; they are simulacra of performances that create their power by other means.

The similarities between the electronic music producer and a band illustrate these means. One day, the band—pianist, guitarist, bassist, and drummer—decides to remix their musical roles. The drummer cuts out the backbeats, turning her attention to the cymbals and keeping time a few beats further afield from a 4/4 meter. The bassist plays long tones instead of locking into the downbeats with the drummer's kick drum. The pianist transposes bass lines into a high register, and the guitarist borrows the drummer's rhythms for sequencing his chords. None of the musicians are playing what they might usually play, yet the music coheres and sounds more interesting than ever.

The band remixing itself illustrates some of the compositional options available to the electronic music producer. While one can aim to meticulously emulate the sound and feel of what live instrumentalists typically do, the tradition of electronic music production is to approach *each sound in the virtual band as endlessly elastic, positioned halfway between the acoustic and the synthetic*. In the producer's sound world, the virtual instrumentalists' sounds may have no connection to the playing histories of their respective instruments: a rhythm part can play melodic tones rather than drum sounds, for instance. The producer Tom Jenkinson (Squarepusher) describes taking this approach in his production work. Jenkinson configures his system so that different instruments/sound types sounds swap their respective roles:

> The thing that fascinates me is more about making a small amount of instruments sound like a lot of instruments, so that sometimes what you thought was a synth is also doing the

drums. I'm stretching things about so that instruments are swapping roles and augmenting each other to build sounds as if they're made out of hundreds of different things.

(Jenkinson, 2015)

As the producer moves around the DAW and manipulates parts and sounds to hear the results, each move frees the producer's virtual band from instrumental conventions. Sounds are free to mix and mingle their roles as necessary, and in fact, the band has dissolved. In the instrumentalist's world of live acoustic music making, there are limits to morph-ability. But in electronic music production, morphing has no limits. The late producer SOPHIE captured this idea, precisely describing a track's ideal state as an "elastic, full frequency range morphing composition" (Kessler, 2014).

In sum, the flexible control the producer exerts over her arrangement, sounds, and mix is *not* a matter of rendering or approximating the sound of a live band. Instead, the production goal is more radical: to improvise the sound of a band without instrumental affiliation or limits on the sounds it can and cannot make, a sound that morphs from one state to another in an instant, a sound that listens to itself on multiple layers of relation and interaction, never tires or becomes bored, and reinvents itself on each track.

Recording maximum musicality with minimum intervention

I'm playing along to the marimba chord sequences. Without a metronome click, I play continuously on the keyboard for several minutes to try to capture something, moving outwards from a center point, from a low register to a high one, from softer to louder, building upon what I played a moment ago by varying a sequence of chords into phrases more elaborate, and then more sparse.

A longer sequence recorded without a metronome click prevents prematurely boxing the music into conventional phrase lengths. In a Reddit AMA in 2018, producer Ryan Lee West advises aspiring producers to work with chord progressions that are longer than 4 bars (or 16 beats in 4/4 time) so as to not lock themselves into the DAW's default 4/4 metrical grid. "I think it helps not to write chord progressions in 4 bars, 8 seems to work fine, but 4 seems to vividly mirror the grid," he says. "I try to write in 3 5 7 8 [bar sequences] etc so that even if the music is obviously 4/4 the chord sequence refreshes this" (West, 2018).

In addition to helping one get lost and un-lost, and freeing oneself of musical conventions by reinventing musical roles for each track, improvising helps the producer capture a sense of maximum musicality with a minimum of after-the-fact intervention through editing, quantizing, or more radical forms of disruption, a topic we return to in Chapter 6. Producers often try imbuing maximum humanness into their music as early as possible in the production process, and playing one's initial parts without any form of correction is a way to preserve this quality. The producer Anton Zaslavski (Zedd) plays parts for his tracks using a keyboard rather than mouse-clicking notes into a sequence for just this reason. "I almost always play things on the keyboard," he says, "because it's faster, and far easier to come up with cool chord progressions than if you try to insert or move individual notes with a mouse" (Tingen, 2017).

Perhaps one reason why improvising captures a sense of maximum musicality is that it allows the producer to engage in fast and intuitive thinking before the slower and considered process of editing the performance comes into play.[4] Guy Sigsworth, renowned for his meticulous productions, distinguishes between fast and slow creativity:

For me, there are two types of creativity: fast and slow. The spontaneous, explosive side, where I'm generating ideas quickly; and the refining, organising side, where I slowly make it fit together into a satisfying whole. It's important to know whether, at any moment, you need to be in fast mode or slow mode.

(Fischer, n.d.)

Mindset

Whether working in fast or slow mode, the producer might experience a variety of mindsets, often at the same time or in quick succession:

casual mind,
focused mind,
adventurous mind,
curious mind,
frustrated mind,
anxious mind,
confused mind,
susceptible mind,
associative mind,
optimistic mind,
humorous mind,
excited mind,
ruminative mind,
neutral mind

At the outset of a project, a producer could experience a frustrated and confused mindset, feeling like there is little to go on, unsure how to proceed. But as we develop the track by trying out various ideas, connecting bits to make larger bits, our mindset may shift from frustrated to curious and associative, and eventually become focused and excited. This shift can happen in minutes, or it can take weeks. A focused and excited mindset leads us to feel optimistic that the track will, eventually, come together, though perhaps not in the way we had imagined it would.

While producing, it can be useful to adopt a neutral mindset that refrains from premature judgement. A way towards maintaining this neutrality is to approach production as a form of problem-solving (DeSantis, 2015). This can be as simple as noticing an uncompelling sound, then trying out various solutions for changing it into something that is compelling. As well, adopting a problem-solving mindset momentarily reframes producing as the pursuit of structure, not feeling. This idea is illustrated by producer Ólafur Arnalds, who describes his production mindset as methodical rather than feeling-oriented:

It's always a little bit methodical to me. When composing, it's not really about the feeling of the music, it's a bit more about mathematics and logistics because you have to put together all these different elements and get a strong structure.

(Meyer-Horn, 2019)

A neutral mindset also allows happy accidents to happen during the production process without our impatience or judgments getting in the way. As discussed in Chapter 1, Brian Eno and Peter Schmidt's Oblique Strategies were aimed towards this way of thinking. Their strategies such as "Look at the order in which you do things" encourage a level-headed reconsideration of wherever one is in a production workflow. When you can, take a moment to ask yourself: *What am I thinking about as I work, and what is the tenor of that thinking?*

Creative outsourcing and resisting the predictable

It is not unusual for electronic music producers to begin making music by letting the complex system that is their assemblage of music software and/or hardware tools suggest directions and lead the way. Producer TJ Hertz (Objekt) relies on his software-based system and his workflows for using it to direct his work. Hertz generates source material by experimenting with processing sounds and adjusting and automating parameters in Ableton Live, without knowing where this experimentation will lead:

> I rely quite heavily on happy accidents or detuning things or processing things in unpredictable ways. Because for me, the results that you get from tweaking a delay line or jabbing at buttons on an unfamiliar synth or recording a result while you automate some parameters where you're not exactly sure what they're going to do is a pretty important part of the process. That's the source material that I work with, and using that as a basis—that's when I can start to let my intuition guide me.
>
> (Martin, 2018)

Similarly, producer Calum MacRae (Lanark Artefax) describes how he begins a track without a specific sense of what he wants to do, preferring to "just let something emerge." He also resists relying on conventional rhythms, meters, or keys:

> I start from a position where I try not to do anything predictable. I prefer to sit down without having an exact idea in my head about what I want to write; I usually just let something emerge. A good way of doing that is to always try and position yourself outside of various rhythms, tempos, time signatures or keys. If you start outside of this you can find something interesting to work with.
>
> (Hinton, 2017)

MacRae often begins with percussion sounds—not to build steady beats necessarily, but to use as building blocks for more abstracted percussive timbres and the track's overall soundworld:

> I reckon that consciously starting with percussion or the kinda building blocks of the music forces you to approach those aspects in the same way you might approach melody or to endow it with the same musical affection or whatever that's conventionally afford[ed] to melody or the seemingly "emotional" aspects of music.
>   Starting out with the seemingly inert aspects of music, kicks, snares etc. and using them to articulate something sorta greater than themselves seems to me to be the easiest way to start unpacking and reassembling your own idea of what music "should" sound like.
>
> (Girou, 2016)

In sum, MacRae spells out a four-part strategy for resisting the predictable that any producer would do well to follow in their pursuit of interesting musical material. First, let ideas emerge without imposing a direction. Second, side-step conventional ways of structuring a piece. Third, find starting points in "the seemingly inert aspects" of your sound palette. Finally, from these decisions and discoveries, build your sense of how the music should sound.

Conclusion

This chapter has outlined my experience beginning *Plentitudes* by sampling my marimba playing and building upon that. It also offered some general concepts that electronic music producers might think about. Pärt asks us to devise the appropriate system for our musical gesture. Cage suggests we discover the dialectics among our ideas and our musical system by listening to how they "talk" to one another. Budd urges us to focus on a small thing to discover everything in it that makes musical sense, while Hopkins thinks in terms of compositional placeholders and through-lines. We also considered how improvisation shapes many facets of electronic music production, from finding interesting sounds, happening upon accidental counterpoint and finding variations, being lost and getting un-lost, emulating the sound of a band, and capturing maximum musicality. Finally, we considered the impact of mindset, creative outsourcing, and steps for resisting the musically predicable. In the next chapter we turn to presets and sound design.

The faintest traces Quest

Ideas are everywhere

A musical idea can be anything: a melody, a chord, a beat, a sample, a field recording, or simply an interesting timbre. Pay attention to any musical presence that seems compelling at this moment and capture it.

Repeat it

Repeating your idea allows you time to listen to it, assess it, and give it an opportunity to work its perceptual magic. Sometimes we do not know how interesting an idea is until getting to know it through repetition.

Vary and develop it

Develop a vocabulary of production moves by which to vary and multiply your ideas.

Notes

1 Argeïphontes Lyre is available at akirarabelais.com. Some of its quirky preset names include "Eviscerator Reanimator" and "The Lobster Quadrille."
2 Motherwell once said: "All we needed was a creative principle, I mean something that would mobilize this capacity to paint in a creative way" (Cummings, 1971).
3 Retroactive recording features in DAW software, such as Live's Capture and Cubase's Retrospective Record, work by recording everything you play, even though you are not in Record mode. When you are finished, click the Capture/Retrospective Record button to recall the MIDI notes you just played. In this author's experience, Capture and Retrospective suggest an ideal workflow in which you play first, then go back later to refine it. In production workflow terms, you were composing the whole time.
4 This idea that there are two modes of thought – one that is fast, intuitive, and emotional (System 1), and another that is slow, deliberate, and logical (System 2) – comes from the work of psychologist Daniel Kahneman (2013).

4 Subliminal feeling

Presets and sound design

Introduction: the Laxity sample controversy

In early 2020, pop megastar Justin Bieber released his album *Changes*, and within 24 hours, the album's second single "Running Over" was accused of sampling the melodic sequence from "Synergy," a 2019 track by a singer named Asher Monroe, and "Flight," a track by the Korean hip hop artist, YUMDDA. As the plagiarism accusations rapidly escalated, a young Zambian producer from the Netherlands named Laxity clarified the situation on Twitter: Bieber hadn't stolen a sample from Monroe or YUMDDA, because all three artists were using the same loop from Splice, an online platform for royalty-free samples and sound packs (Deahl, 2020). The contested loop was part of a Splice pack that Laxity himself had created, uploaded to the platform, and used on his 2018 track, "Good Morning." To prove his authorship, Laxity posted a video showing the original DAW project file on his laptop, including the melody's MIDI and the VST plug-ins he used to process it, noting: "I made the melody, saved it as a sample and released it on @splice for ANY one to use" (Laxity, 2020). A month later, Laxity was interviewed on Splice, where he shared a video tutorial on how he had produced the loop (Splice Blog, 2020).

On Twitter, the Laxity sample controversy sparked discussion about the ethics of using prefabricated samples in electronic music production, and producer-fans weighed in. A Twitter user named Singto Conley argued that "if a producer working for someone as big as Justin Bieber just drops in a splice loop with minimal editing they should at least lend some written credit to the artist who made it, just an honorable gesture imo" (Laxity, 2020). Conley compared the Bieber-Laxity controversy to Lil Nas X's country music-tinged hip hop track, "Old Town Road." When the track became a viral hit in 2019, Lil Nas X had publicly credited another young Dutch producer, YoungKio, whose self-described "throwaway beat" Lil Nas X had found and purchased online (Lamar, 2019).

The controversy over Bieber's production team's use of Laxity's sample illustrates how critical—and noticed!—sounds and sound design are to contemporary music production: the right sound communicates subtleties of style, telegraphs the taste of the producer by situating them within a musical community, and in some cases, can be the x-factor that helps make a track a hit. But even as samples, sample packs, loops, and FX are effortless and affordable to download thanks to Splice and other subscription-based websites, there can be stigma attached to using preset sounds without at least minimal editing or processing to customize them, disguise their origins, and make them more bespoke. Bieber's use of Laxity's sample, even though perfectly legal and an example of a widespread and accepted practice, could be heard as a production short cut.[1]

While the distance between pop stars using Splice loops and this book's focus on the work of more experimentally oriented electronic music producers who craft the materials for their tracks' sound worlds may seem vast, both cases are part of a single production continuum. All electronic music producers, no matter what style of music they make, no matter how popular or underground they are, traffic in the minutiae of sound by paying attention to levels of detail within details that impact the feel of their music, a topic we return to later in the book. This chapter explores matters of sound design in the production process. I examine the production landscape of sound presets, production "composing kits," strategies for limiting one's sound sets, the 80% rule of sound browsing, the acoustic sound ideal, and the uncanny valley aspect of acoustic- versus synthetic-sounding sounds. I conclude with discussion on the importance of sound design as creative process in electronic music production.

Presets

After arranging my marimba chord samples into sequences, I began searching for rather *generic* electronic sounds that might go along with them. My idea is to assemble a *sound set* that will complement and contrast with the mellow timbre of the marimba. Searching for sounds with only their broad timbral contours in mind—soft-attack, middle register pads, a thin bell-like lead, some bright percussion, a rounded bass—I comb through my software instruments and their presets, as well as other sounds I have made myself with these instruments, listening.

Today, preset sounds are accessible and affordable, but earlier in the 20th-century they were novel kinds of musical building blocks. Presets can be traced back to electric organs from the 1950s to the 1970s. After WWII, as organs became widespread in American homes, instrument manufacturers offered one-touch chords (e.g. a C-major chord when the C key is pressed) and rhythm auto-accompaniment presets (e.g. buttons for bossa nova and foxtrot beats) to facilitate playing one-man band arrangements of popular songs at "family sing-alongs, tapping into the market for the popular dances of the time" (Wilson, 2016). One of these organ companies was Chamberlin, founded by an American engineer named Harry Chamberlin. After playing in dance bands in his youth, Chamberlin bought an electronic organ and a tape recorder—in other words, a simple musical system—with which he could make recordings. Chamberlin would later explain how the idea for a tape-sampling keyboard device had occurred to him:

> And I was putting one finger down like this, and I said, "For heaven's sake. If I can put my finger down and get a Hammond organ note, why can't I get a guitar note or trombone note and get that under the keys somehow and be able to play any instrument?" As long as I know how to play the keyboard, I could play any instrument.
>
> (Epand, 1976)

Chamberlin's first sampling instrument was a rhythm machine he built in 1949 called the Chamberlin Rhythmate, a stand-alone large wooden box with a giant speaker, designed to be used to accompany organ players. With no keys or buttons, the Rhythmate had instead a slider for triggering tape loops of 14 different rhythm presets. These presets were loops of acoustic drums playing various rhythms; the loops could be blended together and their tempo adjusted. Ten years after the Rhythmate, the American organ company Wurlitzer released the Side Man, which used tube circuitry and a rotating metal disc rather than tape loops to simulate preset percussion sounds including snare drum, temple blocks, and claves. The instrument's 12 rhythm presets included the beguine, tango, and rumba. Marketed as a "full rhythm section at your side!" that could serve as accompaniment for Wurlitzer's popular electric piano, the Side Man worked well enough that it worried the American Federation of Musicians' Union, who feared that its presets "would be used to displace a live performer" (Saydlowski, 1982: 20).

Just as rhythm presets had been a part of home organs, by the 1970s and 80s other instrumental sound presets were becoming a standard part of synthesizers by Moog, Sequential Circuits, Roland, Korg, PPG Wave, and Yamaha (Théberge, 1997: 75–82). Many of these presets became essential timbres in popular music of the time. A list of prominent examples includes: the Moog Polymoog 280a's "vox humana" preset used on Gary Numan's "Cars," Sequential Circuits Prophet 5's nasal-sounding "sync" preset on The Cars' "Let's Go," PPG Wave's "PPG Choir" preset on Depeche Mode's "See You," Fairlight CMI's "shakuhachi" bamboo flute preset on Peter Gabriel's "Sledgehammer," and its "orch5" orchestral stab sample preset on Afrika Bambaataa and Arthur Baker's "Planet Rock" (Simpson, 2018). The most famous preset of all is "E PIANO," an icy, Fender Rhodes-like electric piano sound from Yamaha's DX-7, an FM synthesis digital synthesizer that was hugely popular in the 1980s. In 1986 alone, "E PIANO" was used on roughly 40% of Billboard's Hot 100 #1 singles (Lavengood, 2019).

From early home organs and rhythm machines to 1980s synthesizers and today's DAW software, presets in electronic music production have become omnipresent, and extend far beyond drum loops and synthesizer sounds. In his insightful book devoted to the topic, producer Stefan Goldmann explains presets' reach within production:

> they have spread to the most unlikely areas of digitization: replications of acoustic instruments being at the verge of acceptance in the context of classical and traditional musics; entire chains of guitar amplification including characteristics of speakers and microphones, saved into digital profiles, already employed in rock and pop production.
>
> (Goldmann, 2015: 10)

To illustrate Goldmann's observation, consider the presets in reverb VST plug-ins such as Valhalla's VintageVerb, Sound Toys' EchoBoy, and Goodhertz's Megaverb (Figure 4.1). These plug-ins include dozens of emulations of classic hardware from the 1970s and 80s. Megaverb, for example, includes a collection of "Historical" reverb presets named after and modeled on sounds from iconic pop and electronic music albums and tracks, including David Bowie's "Low" from 1977, Prince's "Kiss" from 1986, and Aphex Twin's ambient classic, "#3" from 1994.

Perhaps due to their ubiquity, presets have become integral to DAW-based electronic music production. DAWs includes thousands of ready-made sounds and effects available to be used however a producer wishes. Cloud-based music production subscription services such as Splice, Noiiz, and Loopcloud are another route to presets, offering producers "access to large libraries

Figure 4.1 Reverb presets in Goodhertz's Megaverb VST effect

of audio samples, synthesizer presets, VSTs . . . partially completed works, or completed works for remixing" (Shelvock, 2020: 6). Splice alone has several million subscribers who use the platform to source samples for their tracks. And while the example of Justin Bieber's production team using Laxity's Splice loop might be a production short cut, producers of all levels of accomplishment work with presets in one form or another, if only as starting points. As Matthew Shelvock details in *Cloud-Based Music Production*, hip hop producers use Splice to select, manipulate, and arrange samples into their tracks (ibid.: 23-36). Moreover, in hip hop and other styles of electronic music production, it is common creative practice for producers to not only use presets but also create their own *sample packs* as a way of disseminating their work.[2]

A producer's familiarity with presets demonstrates a kind of literacy, notes Alan Durant, "a question of knowing bits of music history . . . rather than knowing anything technical about sound envelopes, filters, or instrument design" (Durant, 1990: 188–189). Using or altering presets also teaches us about sound design. When producer Mark Fell began learning electronic music production as a teen, he worked with a Yamaha TX81Z synthesizer, making his own sounds from the instrument's presets:

> I was designing my own sounds, but often I'd start with the preset and then modify it. There was one cymbal sound on the TX81Z—a metallic, percussion thing with a nice decay on— that I remember just doing endless variations of. So often I'd start with a preset and then modify it or start from scratch. . . . There's always been this negativity around using presets but I have no problem with them.
>
> (Wilson, 2018b)

Producer and educator Tom Cosm, whose production tutorials are explored further in this book's Interlude, explains that the immediate gratification aspect of using preset sounds and effects inspires producers to learn more about sound design. Cosm has created Effects Rack packs for Ableton Live, including a pack called Downgrade, a set of five racks for creating low-fi sounds that use Ableton's stock effects (e.g. Downsampling, Pitch, Saturation, Distortion, and Echo). In a podcast interview with fellow producer Mr. Bill (Bill Day), Cosm explains that he intentionally simplified the GUI of the racks by hiding their effects behind Ableton's "Macro Control" knobs. This way, users could quickly experiment with his presets and stay focused on the production process:

> Cosm: The Downgrade pack . . . were racks for Ableton Live that pretty much consisted only of native Ableton effects but did very low-fi hip hop kind of stuff. I would have a bit of tape wobble and then add some noise in different ways, some band restriction with EQs to make it sound like an old radio or crackles and that kind of stuff. I made sure that all of the devices were hidden and there were just the eight Macro knobs. And that is by far the most popular thing that I've put out.

Cosm's Downgrade pack's hidden devices reminded Mr. Bill of FL Studio's Soundgoodizer, a multiband maximizer plug-in that similarly combines several parameter settings into a single knob:

> Mr. Bill: It's kind of like when you open up a plug-in like Soundgoodizer in Fruity Loops. It's literally just one big knob that you turn up or down that makes your shit sound fatter or less fat. Obviously as you're turning it up its removing certain nasal mid-frequencies,

boosting bass, boosting highs, compression, soft-clipping, saturation—it's doing a lot of stuff as you turn that knob up. But from the end user experience, you don't give a shit: you just put it on your stuff and turn it up.

Cosm: Stuff like that is really, really good . . . If you can get excitement straight away . . . it just drives [music producers] through the boring shit. . . . If you can carefully craft check-points where people are like—*Yeah that's awesome—here's a knob that does cool shit to whatever we've done!*—I think you're onto a winner in terms of keeping people retained.

(Day, 2020a)

Despite the plethora of presets and ways of obtaining them via the DAW, production websites like Splice, or bespoke Effects Racks such as those of Cosm, some producers avoid using presets without significant alteration and customization. For example, producer Laurel Halo describes encountering "the most awful trance, drum & bass, dubstep kind of presets" in her Access Virus hardware synthesizer. Halo's strategy is to transform them into her own original sounds. "There's a way to take these awful presets," she explains, "and maintain that contemporary sheen and edge but make your own sound out of it and make it sound good" (Keeling, 2013).

Another example is the producer Joris Voorn, who admits that he's a "preset guy" who primarily works with VST synthesizer presets he has altered. "I am not someone who will make his own sounds out of a modular system," he says, "or start with one sine wave and create a whole crazy pad." Voorn defends his practice as anti-purist, astutely pointing out that many now-classic electronic music timbres began as presets:

If there's one thing I don't like it's purism. A lot of tracks today go back to the most preset-y sounds you can find, which is basically the [Roland] 909 drum machine. That's the blueprint of techno and it's been a preset since day one. What's the difference between a preset and a guitar? A guitar is a preset as well. There's some really great presets in music, and one of the interesting facts about electronic music—techno especially—is that it has given birth to a few sounds that people really know and are so familiar with. And you're free to use them and create something new with it. You make a totally new sounding track using the same preset they used twenty years ago.

(Voorn, 2016)

In his book *Presets*, Stefan Goldmann speaks for many electronic music production purists when he explains that he has never used presets for reasons of creative integrity. "I had been following an unwritten rule," he says, "that creativity in electronic music means, to a greater of lesser extent, engaging with sound design" (Goldmann, 2015: 12–13). Along similar lines, producer Robert Henke (Monolake) finds presets problematic because he experiences little connection with them. "With presets," Henke observes, "there is no response connecting what I hear with myself" (ibid.: 33). Finally, producer Tom Jenkinson (Squarepusher) hears preset-based electronic music as proof of producers who have not moved beyond the default sound capabilities of their software or hardware instruments:

I do think that we're seeing a lot of music that is preset-driven. I guess sometimes the instrument has an architecture that will steer you around; it will present possibilities that are easier to do than others, but a lot of musicians seem to be demonstrators, because they've basically followed the path of least resistance with an instrument and are therefore exposing its principle characteristics.

(Jenkinson, 2015)

Case study: Spitfire Audio's Evo Grid and Composer Toolkits

Spitfire Audio makes sample libraries, sample-based virtual instruments, and composing "toolkits" aimed at electronic music producers and film composers (whose work increasingly intersects). Spitfire is known for meticulously recorded orchestral libraries of string, brass, woodwinds, and percussion samples that can be loaded into a VST sampler and triggered using a keyboard. These libraries include samples of different playing articulations and techniques (e.g. soft or legato attack, vibrato or clean tone) recorded from different microphone positions. Among Spitfire's innovations are its Evolutions libraries, based on the company's Evo Grid software, and its "Composer Toolkit" collections. The Evo Grid software is modeled on the EMS VCS3, an analog synthesizer from the 1960s with a peg board-style grid interface. The GUI for the Evolutions libraries allows a producer to move pins around an EMS-style, Evo Grid peg board to select various randomized instrument articulations (Figure 4.2). These randomizations produce timbral complexities within long tones that approximate the sound of live string players changing their note articulations over time. Spitfire describes its Evolutions system as an "effective means of easily writing music that is able to subtly change over time without variation in melodic content" (Spitfire Audio, n.d.a). In a way, plotting randomizations on the Evo Grid accomplishes the kinds of subtle and unpredictable variations and complexities that many electronic music producers aim to incorporate into their tracks by other means, a theme we explore in Chapter 6.

Spitfire's Composer Toolkits are based on the sounds and production techniques of cinematically-oriented electronic music producers, such as Ólafur Arnalds (Spitfire Audio, n.d.b). The Ólafur Composer Toolkit is based on samples of Arnalds' felt-muted piano sound recorded through various high-end microphones and signal processors. For example, the Toolkit's Processed Piano sound preset is described as "a mix of the [Neumann] KM84s with a pair of [Telefunken] C12s, mixed by Ólafur into a pair of Pultec EQs, then slammed through a [UREI] 1178

Figure 4.2 Spitfire Audio's Evolutions Evo Grid GUI

stereo compressor to give you the distinctive Arnalds sound" (Spitfire Audio, n.d.b). In addition to the piano sounds, the Toolkit includes 100 "sonic warp" presets that Arnalds created by processing piano and synthesizer sounds into ambient textures. In a review of the software, a producer-fan named Edward Abela explains that, for him, the Toolkit's sounds evoke real electronic instruments and recording spaces: "The main thing I love about all Spitfire's sounds is that the space they build around the sound is absolutely fantastic. It lets you feel like you're in the room with the instrument rather than hearing a sample off of your computer" (Abela, n.d.).

Spitfire's products are positioned at one of the leading edges of the contemporary preset and sound design world, insofar as they combine the realism of acoustic sample sources with options for manipulating those sounds informed by sophisticated musical and recording practices. Moreover, while the libraries may be aimed at film composers and cinematically oriented electronic music producers, their sounds have migrated to electronic music contexts further afield, illustrating "the inter-flow of production trends" (Collins et al., 2013: 105). Consider, for instance, a moment from a YouTube tutorial in which producer Conor Corrigan (Emperor) loads the Ólafur Composer Toolkit into his DAW and explains how he uses it for atmospheric sound design within otherwise intense drum and bass tracks:

> It's obviously really beautiful, but you can use these [Composer's Toolkits] for anything really. . . . Space and stuff. By altering these parameters you've got [in the interface], you get these really nice little movements . . . So, with this you could go in two directions, I'd say. I could make that into an atmosphere, or I could even put this through this bass [effects] chain I was doing earlier . . . So all this kind of stuff you could just stick a recorder on and just record the whole process—you're playing around with the sound, and I find something that I like and just cut that out of the session and take that piece. . . . That [sound] could even be used as a carrying bass line in a drum and bass track.
>
> (Corrigan, 2018)

Sound design

> If I could create my own sound, then that meant I could create music that broke free from existing ideas
>
> —Isao Tomita[3]

Whether one uses presets and sample packs or creates one's own sample-based instruments, sound design is an ever-present and integral part of producing electronic music. The reason for this is that producers want control over the timbral subtleties of their sounds: such subtleties communicate reams of meaning, and small changes to a sound can have profound perceptual effects. Producers' sound design also reveals their experience with their tools and their skills at listening closely to sounds—that is, attending to micro-differences that make a difference in the overall feel of a track's production. As the rapper Big Sean once said about the beats of producer Mike Will (Mike WiLL Made-It), "You can tell from the textures somebody took the time to make sure that the sonics are right" (Seabrook, 2016).

The significance of sound design explains why there are so many production tutorials online that focus on the topic and show how to achieve a specific timbre or production effect in a DAW. For example, videos on producer Thomas Bevan's SynthHacker YouTube channel explain how to build sounds in Serum, a VST wavetable synthesizer, similar to the "organic"-sounding pads used by future bass producers Flume and Louis The Child (Bevan, 2018). Or taking a more open-ended approach, a tutorial by the producer Nolan Petruska (Frequent) explains how to use Serum to create growling-morphing bass sounds for contemporary bass music (Petruska, 2017). A third sound design example comes from Ableton. On its website, the company offers tutorials on how to create sample instruments from an audio file loaded into Live's Sampler device. In "How To Create Your Own Instruments," the producer Ransby samples a single piano note and creates pad sounds from it. He reminds viewers that while sound design is a process with uncertain outcomes, it is always a learning opportunity, no matter what happens. "Maybe [the sound] is just going to be useless and that's fine because when you're experimenting, you will fail," he says. "Then again, what does it mean to fail because I think I'm learning from everything I'm doing" (Ransby, 2020). Ransby is right: for most producers, sound design is a never-ending process of small failures that occasionally leads to discovering fascinating timbres.

For producers who work with samples, sound design is woven into their workflow. Producer Amon Tobin speaks of processing samples to isolate specific frequencies within them. "You can take out an entire frequency that holds an instrument, so that you can no longer hear it," he says, "or you can hear it in such a background way that it becomes an interesting subliminal part" (Young, 2003). Producers often manipulate a sample's timbral and rhythmic profile until the sample is original. For example, producer Bryan Müller (Skee Mask) manipulates his samples so that they become his own:

> I try to make everything myself. Even when I sample, I really try to program with the sample so that I'm not just reveling in the value of the sample, but instead I approach it as simple sonic fabric that I then basically re-stitch completely so that it doesn't sound all the way that it did before rhythmically. I do try somehow to breath [sic] new life into it from my end, so that it really does become like my own instrument, as soon as I can put certain effects on it or when I run it through my effects units.
>
> (Müller, 2017)

Searching for sounds and committing to a few of them

As I audition sounds from my software instruments, I have a sense within a few seconds whether or not they could work in the project-in-progress. I tend to avoid sound FX, and also hyper-unrealistic sounds, although in the right context, every sound has its use. I seek sounds with a built-in sense of mystery that I will enjoy working with throughout the production project and, hopefully, beyond. Oftentimes, this means identifying or creating sounds that have layers of harmonics, noises, or other ambiguities built into them.

At this stage, however, I need only a few generic timbres because over time the sounds I chose will probably change—I might even swap them out for other sounds. Making do with just a few generic timbres simplifies the production process and focuses our attention on how to do things *with* and *to* the sounds rather than doubt whether or not they are the perfect sounds for our project. Since no sound need be forever fixed in the digital realm, the producer does not need to spend excessive time at the outset of building a track finding perfect sounds. Even the simplest projects will gradually become more complex as the producer devises ways to turn generic timbres into more compelling, original presences. Our sound sets inexorably evolve as we uncover their potential to become other than what they are right now. As TJ Hertz (Objekt) says about his process:

> I rarely end up where I set out to. Almost every track I write goes through at least three or four major phases, which in some cases can sound like completely different tracks, over the course of 40 to 80 versions.
>
> (Smith, 2018b)

I settle on a sound set consisting of two analog synthesizer-type pad sounds, a sine tone + white noise bell sound, two drum kits, and a sub bass. These classic (if not particularly exciting) sounds are enough of a palette with which to play along to the marimba samples. I try out sounds and parts, layering the different timbres until I have bits of call and response dialogue happening through chords, melodies, and rhythms.

On music production disconnects

Pondering the potentials of my sounds that can change over time and seeking immediate expressiveness by using those sounds feel like two different ways of being. Since my musical system is software-based, it is by nature always more open-ended than I can fathom. Where exactly, for example, does a VST synthesizer *end*? With the 400 presets that come with it? With the sounds I have made with it? *There is no end to software's sounds* because (1) we can change sounds inside the instrument and (2) we can change its sounds externally by routing them into other plug-ins inside the DAW's ecosystem, or beyond it. We ponder the open-endedness of our tools: *What will happen if I route that sound into this device?* But the horizon of our musical system always extends further than we can go, promising ever more enchanting sounds . . . if only we could try just one more combination.

In contrast to the open-endedness of my software, my production workflow is goal-oriented and now-focused. I use whatever is closest at hand—*such as Pad 35 I made last week*—and try to make something from it to hear what happens. Trying things, I assess and adjust, making annoying things less so:

> that sounds too dark—lighten it up; it's muddy, make it clearer

Such impressions happen in quick succession, and I attend to their urgency whilst trying not to get caught up in their demands, to keep the production process moving. Once in a while, the disconnect between potentials and expressiveness, technical tinkering and feeling, disappears and my software tools seem *alive*. For a moment, they are responsive to exactly what I want to do. There is no software, no screen, no speakers. I forget I am making highly mediated music: the computer has vanished and the sounds are *expressive*.

Frames for attention: limiting one's sound set

One of the producer's primary tasks is figuring out what materials to work with, which helps focus attention. Since our process won't fully reveal itself until we're further along, there is no sense worrying about that until we get there. But our materials we can decide upon now, even if that deciding feels arbitrary because *one's materials are perpetually in flux*. To negotiate this prospect of perpetual flux, producers find it helpful to work with a somewhat limited set of sounds. Drew Daniels, of the production duo Matmos, suggests committing to a single sound source:

> It's easier to focus when you make a strong commitment to one concept or one sound source or one idea and then follow through in a really fanatical way. But you have to be careful what you commit to: once you're inside the framework, it's like a dare that you've made and honor demands that you must carry it out.
>
> (Anthony, 2019)

Producer Tim Hecker also commits to a constraint on sounds:

> I have 10 kinds of pianos on my computer. I have a real piano I can use. I am just lost with infinite choices and it's like more and more of a problem, where constraint really matters, and omitting really matters, and putting on horse blinders really matters.
>
> (Kretowicz, 2016)

The prospect of being lost with infinite choices spurs producer Beatrice Dillon to limit herself to just a few elements:

> When you switch on your computer, for the first ten minutes you're buzzing and then you realize there are too many options and hit a wall. . . . I'll usually say, three colors, go . . . any constraint to get me going. Then you get really attached to those few elements and look to do more with them rather than piling on elements. I like people who do a lot with a few elements.
>
> (Turner, 2020b)

Beginning with a single set of sounds does not mean that their attributes will not change over the course of a production project. It is inevitable that the sounds will change in ways both small and large as we develop our tracks. Since his productions undergo dozens of iterations, TJ Hertz explains that it is not especially important which drum sounds he begins with, because they will eventually be replaced:

> Given that all of my finished tracks sound almost nothing like how they did when they started, I guess ultimately it's not really that relevant how I begin. I guess usually drums, but they rarely last more than a few versions before being replaced anyway.
>
> (Hertz, 2016)

The producer Jerrilynn Patton (Jlin) concurs:

> Most times the track is completely unrecognizable by the time it's finished. The only thing I might keep from an initial idea would be a set of hi-hats or something like that.
>
> (Electronic Beats, 2016)

Assembling and working with a *cohesive sound set* is also key. Like Matmos, producer Calum MacRae (Lanark Artefax) works with a single set of related sounds, manipulating them in different ways on different tracks: "I generally work from the same bank of sounds and I have these digital instruments that work organically so as to do very different things each time I use them" (Eede, 2017). Producer Scott Morgan (Loscil) likewise works with field recordings of soundscapes and instruments, "building a library of sounds to work from first before entering into composition mode" (Headphone Commute, 2014). With a library of samples, Morgan begins "manipulating them and building textures out of those found sounds that kind of get reassembled into a fairly dense and layered deep listening experience" (Joe M, 2019).

The duo Boards of Canada, who have been influential electronic music producers since the 1990s, create sample instruments by recording live instruments, applying effects, and resampling the results, a topic discussed further in Chapter 6. Member Mike Sandison outlines their process:

> One technique we like is to create entirely new instruments by sampling ourselves performing on real instruments and then destroying the sounds. So we'll maybe spend days just playing various things, wind instruments, strings, guitars, bass, synths, for hours into the samplers and then feeding those sounds through stacks of destructive hardware and resampling them to make unrecognizable new sounds. This is all before we even begin writing any tunes.
>
> (Pareles, 2013)

The workflow of producer Phil Robinson (Philth) relies on pre-production time spent making drum and bass samples to use in his tracks:

> Sometimes I will just spend two hours doing bass, and that's it. And what I'll do is export all of it, and what you get is a library of basses . . . I do pre-production sessions where I just do drums, and then stop.
>
> (Robinson, 2019)

"Workflow," Robinson reminds us, "is being able to work fast" (ibid.).

Like Morgan, Boards of Canada, and Robinson, producer Alan Myson (Ital Tek) also amasses a library of homemade sounds, including live percussion and singing. For his 2016 album *Hollowed*, Myson describes his process as *just experiments* for designing its sounds:

> A huge amount of time was spent on sound design and just creating the sonic world for what the album could be. I also need to give myself the tools to make the album, so I give myself a lot of time recording stuff. I've got a whole selection of recordings of different instruments, making sample patches and recording vocalists. Recording my own percussion in weird spaces just to give myself some sources to be inspired by. It's all just experiments, and I just see what works, and the vast majority don't, but when I see a path that leads somewhere, I just hammer down on it. If I have enabled myself to have enough tools, when that moment strikes, I can just go with it.
>
> (Carr, 2018)

The producer Mr. Bill separates some sound designing from his sequencing, editing, and arrangement work. Bill explains how he incorporates what he calls *sound design mud pies* into his tracks:

> What makes a song interesting . . . is pretty much all the sound design and editing stuff. A mud pie is a long file of sound design. . . . It's just a bunch of noise, basically. But then with careful editing and interesting effects, you can sort of chop all of that up to make a riff.
>
> (Pyramind, 2020)

Finally, the producer Greg Jones (G Jones) recommends that producers constantly experiment with sound design, without fear of making unusable sounds, in search of *magical combinations* created by unexpected interactions within one's musical system:

> My best advice is to not be afraid to make a bunch of horrible noise in pursuit of something special. The more you f--k around and try new things that seem like they won't work, the more likely you are to discover some magical combination of sounds or effects or whatever

that you would not have discovered if you were trying to achieve a specific sound as opposed to having an exploratory approach.

(Kraker, 2018)

The 80% lesson

Having a library of presets and sound design—pre-made, customized, or bespoke—at the ready helps the electronic music producer work quickly. As Phil Robinson says, *workflow is being able to work fast.* It is not unusual for producers to have thousands of ready-to-use sounds on their computers and then search for something on the fly that fits their track. But some producers require an even lower-friction way of proceeding to safeguard against getting hung up on sound design minutiae too early in the production process. Steve Duda, the developer of Serum, explains what he calls the *80% lesson,* which he learned from watching producer Joel Zimmerman (deadmau5) producing. As Duda explains, Zimmerman doesn't obsess over the perfect sound, which keeps his workflow moving quickly:

> Joel's taught me a lot about being able to hear something in your head . . . you go for it, and when it's close, you move onto another facet. You don't obsess over this one detail, because if you obsess over this one detail you're going to lose the big picture. So that was something he never verbalized to me, but just working with him and watching him work is something that was really valuable to see. It's really easy with all these tools to get distracted. . . . The more that the music comes from within you, the more that you know what you're going for, the easier it is to find it. . . . Once I sat down with him and we were working on music together, I saw certain habits from him. . . . He had some really weird things that he did that I had never seen people do before. One of them was to just fly through samples. If we were looking for a kick drum, it was just [mimes quickly hitting a key on the keyboard] *next, next, next, next, next . . .*
>
> It took me a while to realize . . . in his head he already knew the sound of the kick drum that he wanted and he just was waiting to find the one that was closest of a match. And when he found one that was close—*yep, that will do*—and threw it in there, and you're done with the kick drum—now we're onto the next thing. Not, *let's obsess with putting some compressors and some EQ* . . . no, just throw that in there and it's close enough and you're moving on. And if you repeat that with your claps and your snare and your [hi] hat and your bass and your lead and all that stuff—there's something—I think kind of magic happens. Even though each [sound] may be only 80 percent of your vision and it's close enough, the synergy that forms from all of these different elements doing 80 percent of what you wanted them to do—you could start to really hear the intent, it just starts to percolate through.

(Duda, 2015)

Tinkering with sounds

The electronic music producer tinkers with sounds, applying signal and effects processing to conventional timbres to make them less so. In a 2018 AMA on Reddit, a producer-fan of Ryan Lee West (Rival Consoles) named dornbirn asks a question about sound design:

> You manage to pull off one of the most organic sound palettes in the electronic game. Your synths seem alive. It's a bit of a lofty question, but how do you pull that off?

(West, 2018)

West explains that he begins with single synthesizer patch that sounds sufficiently compelling on its own, or what he calls "not too plastic." He records himself playing this sound live, "with some performance of the filter, volume, LFOs changing etc on the synth itself." Then he processes the live sound further with effects inside the DAW:

> I try to find one more powerful synth sound that can carry the weight of attention and that doesn't sound too plastic. this is achieved by several small amounts of effects sometimes, delays, reverbs, distortion, slight granular stuttering to break up the signal. repitching up and down an octave to introduce artifacts, many things and there is no 1 recipe always works.
>
> (West, 2018)

On Reddit, West reminds producer-fans that even "the best synths ever made will often sound dull and uninspired if they are with 0 effects." As a way forward making original sounds, he recommends that producers try creating their own custom effects chains:

> creating effects chains of reverbs, delays, distortion, filtering, compression and passing sounds into different versions of these chains. coz something boring and plain can become very interesting through the right chain.
>
> (West, 2018)

Finally, echoing the earlier ideas of Brian Eno, Les Paul, and Joe Meek on production, producer Steve Hauschildt thinks of sound design effects processing and the studio itself not as techniques but as instruments that are an integral part of his musical system for creating unique and multi-dimensional timbres:

> I think it was important to think about effects as being an instrument or a studio as being an instrument as opposed to just doing synthesis on because I think that a lot of people are really into synthesizers and do really nice things with them but I think the palette you get with that is kind of one or two dimensional. So, what I was doing took other steps and other processes . . . my thinking was to bring in other studio's rack equipment, their instruments, third party plug-ins, all of these different [sic] that end up contributing to what my music sounds like by the time someone hears it.
>
> (swc, 2017)

Synthesis thinking

One difference between composing for acoustic instruments and the panoply of virtual and synthetic timbres available to the electronic music producer via the DAW and VST plug-ins has to do with expectation. Consider the piano. When I play piano, I have a reliable idea how the instrument will sound ahead of time and how it will respond to my touch. The layout of the keyboard is familiar and I have a repertoire of go-to moves, chord shapes, and hand positions I can use to build music. I know what to expect from the piano in terms of touch sensitivity, sound, and layout. What makes the piano a piano is always present in the instrument and in my body knowledge of playing it.[4]

But when I sit down to play synthesized or sampled tones, triggering them via a keyboard or pad controller, my expectations are suspended because I have neither a precise idea of what to expect before I begin exploring, nor a sense of where that exploring might end. I can alter a sound by rolling off some of its frequencies, changing its attack and release, modulating or swapping out its waveforms, resampling it, and so on (and on and on), routing the sound through processing and processes until it has become an altogether different presence. Mr. Bill summarizes this open-ended quality of production with advice applicable to any kind of sound designing: "Start with any input you like, and if you process it enough it will sound insane" (Day, 2019). With a few adjustments, a piano sound becomes a liquid bass, a screeching decaying bell, or a rain of white noise. How did I get *that* sound to become this (insane) sound? Thus it is that in electronic music production, any sound can become any other.

The acoustic sound ideal: sounds that sound live/alive, sounds that do something over time

One longstanding fact about many electronic sounds is that they are, generally speaking, timbrally unchanging and so do not always feel alive. The producer finds ways around this by crafting changes to such sounds that are engaging over time. Here, consider acoustic sounds as a model. The sound of a singing bowl, for example, is compelling because it changes over time rather than remaining static (Figure 4.3). A struck bowl produces a sound whose amplitude rises and falls—a sharp attack followed by a very gradual decay into nothingness. Moreover, acoustic sounds are infinitely responsive to the micro variations of our touch. Each strike of the

Figure 4.3 Nepalese singing bowl

bowl sounds different, as tiny changes in my playing alters the sound. This close connection between playing and sounding is one that many electronic music producers emulate or evoke—consciously or not—through close attention to sound choices and sound design in the pursuit of "movement and dynamism" (Future Music, 2017). In his book, *Composing Electronic Music*, Curtis Roads suggests shaping each sound so that it resembles the changing quality of an acoustic performance:

> In the real world of acoustic instruments, every note and drum stroke is unique. Thus we seek to avoid the bland impression of repeated sounds that never change. It is worthwhile to take the trouble to articulate a unique identity for each object by creative editing and processing.
>
> (Roads, 2015)

A hybrid approach is to combine the best of both the acoustic and electronic/synthetic worlds by using, for example, samples of acoustic sounds and devising ways of editing electronic sounds so that they sound less so. The producer might design sounds by recording live instruments and turning those into sample instruments. This way of approaching the acoustic/synthetic divide enables the creation of an ambiguous timbral palette. On his 2016 recordings, *Gravity* and *Spells*, Ben Lukas Boysen reworked recorded strings "into pads and plain atmosphere layers beyond recognition." He expresses the excitement felt by many producers when he says, "I love turning a very direct and handmade source into an almost metaphysical new thing—that's where digital tools of any kind can be extremely helpful" (Headphone Commute, 2016). Sometimes Boysen's sounds, which appear to be acoustic live performances, are in fact sampled and then meticulously edited to sound live. He explains that, to make the piano parts on *Gravity* and *Spells*, he used piano sample libraries rather than a recording of live piano playing: "I did not record the piano at all but utilized very well made libraries. I can play the basic chords of most tracks but to give the songs some of the human touch back I have to arrange and edit the piano takes a lot" (ibid.). Whether he begins with live acoustic or programmed electronic sounds, Boysen aims for a mix of both, blending them: "I tried to oppose digital and analogue as good as possible, which means, most digital instruments would get an analog treatment within the mix and many instrumental parts got the full digital effect treatment" (ibid.).

Another approach to combining acoustic and synthetic sounds is to devise ways of morphing them into hybrids with attributes of both. The producer Guy Sigsworth uses software that helps him achieve sounds that are halfway between the acoustic and the synthetic. He describes his production aesthetic as "placing sounds on the boundary of the real and the unreal; between the acoustic and the acousmatic."[5] With the goal of defamiliarizing sounds and complexifying our perception of them, Sigsworth likes plug-ins that make sounds ambiguous and facilitates morphing their timbres over time.[6] The producer uses software that processes audio into its spectral components, as well as physical modeling instruments such as Modartt's Pianoteq, a piano emulation. "A modelled piano in Pianoteq may not emulate an acoustic piano perfectly," Sigsworth notes, "but it can be made to morph in real time in ways no acoustic piano ever could" (Fischer, n.d.).[7]

Boysen and Sigsworth are not alone in their interest in acoustic-synthetic hybrids. Producer Jamie Lidell describes the joys of manipulating the "really complicated artifacts" in audio samples, noting how "that mesh between digital manipulation of acoustic sounds will always keep you fascinated in sound in a way that using something that comes from an analog or digital source rarely gives me the same emotional interest. It's the chaos, isn't it?" (Lidell, 2020). Similarly, producer Alan Myson (Ital Tek) manipulates acoustic sounds to sound electronic, finding magic in what results: "There's something magical about an acoustic sound that's been manipulated electronically to sound like it's a synth. There's so many random harmonics and elements of chance and every time you pluck a string it's different" (Turner, 2020a).

Sonic uncanny valleys

Alternately, the producer can ignore acoustic sound as a model and instead showcase a sound's artificiality, foregrounding its synthetic sonic Otherness. This Otherness raises issues of representation that are also at play in digital animation, to which we can take a brief detour to illustrate. Critic Lawrence Weschler writes about the work of programmers and animators at Pixar who digitally render the human form. Their design challenge is to realistically capture the complexities of a person's face, whose many small muscles are deployed in thousands of ways in even the simplest of expressions. Animators also have to render skin texture and "the subtlety and complexity of the way light radiates out from the inside" of the face (Weschler, 2012: 5). In their subtleties and complexities, faces seem to radiate our consciousness. But consciousness is difficult to simulate, because viewers are exacting when assessing the realism or fakeness of digital models: we have a tolerance threshold for the simulated real, and when something looks real-ish but is slightly off, the effect is creepy. Here, Weschler draws on *The Buddha In The Robot*, a 1981 book by robot engineer Masahiro Mori. Mori coined the term "uncanny valley" to describe the small yet significant gap between the simulated/machine entity and the real/living entity. The uncanny valley concept explained how "when a replicant's almost completely human, the slightest variance, the 1 percent that's not quite right, suddenly looms up enormously, rendering the entire effect somehow creepy and monstrously alien" (Weschler, 2012: 15). The goal of digital animators then, is to transcend the monstrously alien and move their work ever closer to appearing real.

In their Quest to render consciousness rather than a monstrously alien effect, the work of digital animators is uncannily similar to that of electronic music producers, and the uncanny valley concept illustrates analogous concerns in production. One could make the case, for example, that since the 1980s successive waves of digital musical instruments (as well as more powerful computer processing) have promised sounds that feel less cold/simulated and more warm/realistic than their predecessors. Over that time, many producers have worked to make their music not have "the 1 percent that's not quite right" sound, but rather a sound ever closer to feeling real, a sound that radiates the subtlety and complexity long associated with acoustic music making.

The technologist and musician Jaron Lanier blames MIDI, the digital protocol for connecting music electronics developed in 1983, with reducing music's range of expression. In his book, *You Are Not A Gadget*, Lanier cites the longevity and pervasiveness of MIDI as an example of a technological "lock-in" that has made it difficult for musicians to imagine other, better technological alternatives. Before MIDI, timbre in music was expansive: a single musical note was a limitless field of possibility rather than a data message containing a limited amount of pitch, timing, and velocity information. When a musician plays an acoustic instrument or sings, no two notes ever sound the same, whereas MIDI reduced music's expressiveness and producers embraced this reduction, making electronic music whose timbral range was more limited than that of acoustic music. The problem with MIDI, says Lanier, is that it "squeezes musical expression through a limiting model of the actions of keys on a musical keyboard" (Lanier, 2010: 9). All this may be true, but perhaps Lanier's critique overlooks the astonishing range of non-acoustic expressiveness in electronic music production that sidesteps uncanny valley concerns. This expressiveness is exemplified in tracks such as Sigsworth's "North" for solo piano, which teeters between sounding real and hyperreal, or Autechre's "32a_reflected," whose singing-bowl-like-but-not-quite overtones hover in suspended drone forms that keep resolving, without identifiable chords.

Conclusion: sound design and finding one's production voice

From Justin Bieber's use of a Splice sample to making sounds to accompany marimba samples, this chapter has examined issues of presets and sound design in electronic music production. Today, there are more and a greater variety of sounds available to electronic music producers than ever before. This raises the question: Do producers even *need* to make sounds from the ground up, or should they simply alter existing presets, samples, and loops? Many producers use preset and pre-made sounds to build their tracks. This kind of production is not necessarily uncreative, as it still requires one to make decisions about textural balance, groove, arrangement, and so on. But is such a drag and drop workflow *adventurous* production? After all, the most creative producers exponentially widen their sonic horizons and take their music in uncharted directions by designing their own sounds, a practice that has no end.

We conclude with advice from producer and film composer Tom Holkenborg (Junkie XL), who suggests that producers dive into sound design through a three-step process of listening, recognizing sounds, and then trying to make their own sounds with the ultimate goal of building a custom aesthetic world uniquely fitted to their music (which recalls Arvo Part's idea, discussed in Chapter 3, that our compositional task is to find the appropriate system for our musical gesture). On his YouTube channel, Holkenborg explains the connection between sound designing and finding one's own production voice:

> *Listen to real sound design music*, like Aphex Twin or Jon Hopkins, great electronic artists who use sound design as their major inspiration for making music. . . . Start recognizing sounds. Start recognizing how layers are used. Is it more reverb in the front? More in the back? How many layers do I recognize?
>
> *Try to make music with existing sound design libraries*. Get your feet wet with, for instance, a plug-in like [Spectrasonics'] Omnisphere, which has a lot of sound design instruments in there.
>
> Once you get comfortable making music with existing sound design, then the third step is, *How do I make my own sound design?* Sound design means something that is designed, but sound is always designed for a certain purpose. If you are the composer, you know better than anybody else what that purpose of sound needs to be. Therefore, ultimately you need to design your own sounds. In other words, every composer needs to be a sound designer to effectively use that in their music. Because what you don't want is playing a sound design sound from an existing library and it's forcing you now to write [a certain style of] music. . . . *You need to design a sound that will fit your composition.*
>
> (Holkenborg, 2020)

Sound design Quest

Begin anywhere by making your own sounds.
Warp, manipulate, and morph sounds to create variations of these sounds.
Build a library of your sounds and their variations.
When you encounter a preset you like, alter it and save it.
Create your own "composing kits" based on your sound design.
Build upon your sound design, feed your process back onto itself, and develop more sounds.

Select between three and five sounds from the many that you have in your DAW or have made yourself, and *limit yourself to using only these sounds in a track*.

(Keep in mind that you can change these sounds and that over time they will reveal new qualities.)

Notes

1 This view echoes earlier discourses in popular music criticism. As Jeremy Gilbert and Ewan Pearson note: "It is sometimes claimed by musicians, critics and fans alike that *the presence of a synthesizer or computer in the arsenal of a band or producer downgrades the ontological status of the music they produce*" (Gilbert & Pearson, 1999: 112; italics added).

2 Sample packs are collections of presets with which to make music in a particular style. Some packs capture the sound of a well-known producer, while others explore the capabilities of a single sound source. For example, the YouTube channel SynthHacker, which specializes in tutorials for Xfer Records' Serum, a VST wavetable synthesizer, sells the "SynthHacker Serum Bundle," a collection of 500 Serum presets.

3 (Hara, 2014).

4 In his lucid phenomenology of learning to improvise jazz piano, *Ways Of The Hand*, David Sudnow speaks of the piano-playing hands having an assessing capacity – the hand "takes a bearing," "appraising" and "appreciating" the topography of the keyboard (Sudnow, 2001: 52). As they move about the piano keys, the hands come to know their environment, "aiming for sounding spots" (ibid.: 40).

5 This aesthetic was inspired by Russian literary theorist Viktor Shklovsky's concept of "defamiliarisation" or "estrangement." In his 1917 essay, "Art, as device," Shklovsky writes that "the device of art is the 'enstrangement' of things and the complication of the form, which increases the duration and complexity of perception, as the process of perception is, in art, an end in itself and must be prolonged" (Berlina, 2015: 162).

6 For example, Zynaptiq's *Morph* plug-in. The company describes the software as "a real-time plug-in for structural audio morphing, the sonic equivalent of one object slowly changing its shape to become a different object – for example, one face becoming another" (Zynaptic, n.d.).

7 A perfect illustration of this morphing can be heard in Modartt's promotional video for Pianoteq 7's Acoustic Morphing technology (Modartt, 2020). In the video we hear Bach's Prelude in C major from *The Well-Tempered Clavier* played as its sound seamlessly changes from that of a 17th-century harpsichord, then a 19th-century square piano, and, finally, becomes a contemporary grand piano.

Interlude
YouTube electronic music production tutorials

Introduction: fandom and electronic music production

A theme running through this book is the relationship between electronic music production and electronic music fandom. Fans share with fellow fans their fascination with the object of their fandom—for example, a celebrity, a sport, a TV show, or a musical style—around which their interests coalesce. In the context of the online/social app mediascape, fandom often shows deep preoccupation or obsession, a focus on details and information gleaned from extratextual sources, and a sharing of what one knows and loves with other fans within a web of social connection, appropriation, and performance (Duffet, 2013: 32-33). In short, fandom transforms "spectatorial culture into participatory culture" (Jenkins, 2006: 41).

From its origins in the academic field of media fan studies (inspired by Henry Jenkins' 1992 book, *Textural Poachers*), fandom connects us to musical practice because it helps us to understand music's social impact "beyond formalist concerns" (Punathambekar, 2007: 207) via its fans who "form special, sustained attachments to musical performers or genres" (Cavicchi, 1998: vii). Fandom is also "the essential background for the more 'professional' musical world" (Finnegan, 2007: 18) in which distinctions between "amateur" and "professional" can blur. After all, some music fans are amateur musicians themselves, and their fandom is expressed in, for example, their "obsessive music listening" that guides learning about their craft (Firth, 1996: 55). Fandom equally inspires imitative amateur music making (Brett, 2015), suggesting that one measure of a musician's or band's impact is the curiosity their music inspires among its listeners.

Electronic music production is informed by a subset of music fans who are themselves active music producers. The work of these fans, who throughout this book I refer to as *producer-fans,* is found online in their talk: in the comments and questions they post on YouTube, production websites such as Gearspace and KVR Audio, Reddit discussion threads, and music software user forums. This online fandom often takes the form of a perceptive probing of production workflows and sound design minutiae. This probing shows a convergence of private and public spaces, since the Internet renders fan activity exponentially more public, putting "the distribution of media content into the hands of ordinary computer users" (Duffet, 2013: 236). What is heuristically rich about this fandom is how it exists in a symbiotic relationship with the practices of electronic music production and the work of some of its celebrated practitioners. Producer-fans critique and debate their favorite tracks, compare them to other, similar musics, speculate on the equipment and techniques used to make them, and try to emulate their signature sounds in their own productions. For these reasons, we can understand producer-fans' fandom as a collective sleuthing in search of music's most noteworthy sonic and production attributes (Brett, 2015, 2020).

YouTube electronic music production tutorials

Electronic music producer-fans have many options for sharing their work online, most significantly YouTube, which has become an immense public platform for "peer to peer" music education (Miller, 2012: 184) where musicians of varied skill and accomplishment share videos of themselves making music. YouTube is the richest source of information about electronic music production. From popular channels such as FACT Magazine's Against The Clock, Electronic Beats, Resident Advisor, The Boiler Room, Pyramind, Cymatics, and Sonic State, to a plethora of tutorials and gear reviews by amateur vloggers, professional producers, and online production schools, YouTube gives us a sense of the totality of the electronic music production scene (Brett, 2019). In video tutorials about how to design classic or contemporary sounds, beat programming, arranging and mixing, as well as walkthroughs of new software and hardware by the companies that made them, seemingly every facet of electronic music production experience is explored. The themes of many videos overlap, showing key features, functionalities, and techniques, and highlighting shared obsessions among different users of the same product. For example, ways of using the DAW Ableton Live and Serum, a popular VST synthesizer, are explained on dozens of YouTube channels. In sum, the power of YouTube videos is that they allow us to see, to some degree, how other musicians work, thus giving us a sense of what is possible to do with the technologies of electronic music production.

Propelled by YouTube, the practices of niche amateur music production worlds that were once relatively unknown are now globally accessible. In 1989, Ruth Finnegan wrote about the many forms of "hidden" music making in the context of an English town, exploring music made by amateurs and promoted within their small and local communities. Finnegan suggested the notion of *pathways* as a metaphor for understanding these forms of musical practice:

> These "pathways" . . . avoid the misleading overtones of concreteness, stability, boundedness and comprehensiveness associated with the term "world." "Pathways" also reminds us of the part-time nature of much local music-making (people follow many pathways concurrently, and leave or return as they choose throughout their lives), of the overlapping and intersecting nature of different musical traditions, and of the purposive and dynamic nature of established musical practices.
>
> (Finnegan, 2007: 305–306)

Finnegan's pathway model continues to resonate in the Internet era, as ever-changing tacit knowledge about electronic music production and techniques for making it circulate through overlapping and intersecting online networks, moving dynamically across convergence points rather than remaining isolated within locally bounded musical communities. With a regimen of YouTube tutorials, dedicated producers can learn deeply about principles of sound design, programming, and mixing in the DAW, for instance.

For amateur and professional producers alike, YouTube is a way to make public the usually hidden labor of making music by sharing the workflows and know-how that guide production work. Producers' tutorials also illustrate what Henry Jenkins calls *convergence culture*, "where old and new media collide, where grassroots and corporate media intersect, where the power of the media producer and the power of the consumer interact in unpredictable ways" (Jenkins, 2006: 2). Jenkins suggests that in this convergence, "the circulation of media content . . . depends heavily on consumers' active participation" (ibid.: 3). This is precisely how electronic music producer-fans are critical to understanding the trajectory of electronic music production practices. Convergence cultures, says Jenkins, are interactive spaces "in which fans and other consumers [are] invited to actively participate in the creation and circulation of new content" (ibid.: 290). From a convergence culture perspective, producers who make YouTube tutorials or stream their producing sessions are uniquely situated fans who "connect information, pool resources, express interpretations, circulate creative expressions and compare value systems" (Duffet, 2013: 243). Producers use their videos to share their musical ideas, amplify already known ones, and explore "new trails of discovery, increasing their exposure to performances and adding to their personal knowledge" (ibid.: 237), while at the same time "potentially open up esoteric knowledges to wider populations" (Prior, 2018: 88). Music production tutorials, then, show a convergence of producers' multiple goals: chronicling the composition of new tracks or re-creating elements of old ones, demonstrating and explaining production techniques and problem-solving skills, and reaching a community of producer-fans who in turn provide feedback on the videos in the comments section. As the producers work, we might even say they are *performing producing* as they move their cursors around their computer screens, drawing our attention along with them. Most importantly, their videos demonstrate real time workflows for using a musical system to make music. With this theoretical context in mind, the case studies in this Interlude consider how the YouTube channels of Tom Cosm, Andrew Huang, Bill Day (Mr. Bill), Joel Zimmerman (deadmau5), and Christian Valentin Brunn (Virtual Riot) present knowledge in their production tutorials.

Five case studies

Tom Cosm: "An Introduction to Digital Audio Production"

New Zealand producer Tom Cosm, whose Downgrade Effects Rack is discussed in Chapter 4, was one of the first producers to make tutorials for YouTube. In 2009 he began "An Introduction to Digital Audio Production," a 10-hour, 10-part series aimed at teaching novice producers "how to write electronic music from scratch." The videos guide the viewer through building and mixing a track in Ableton Live. "Everything is explained in depth as I construct a tune while you watch," Cosm explains in the series notes. "I talk to you like you're sitting next to me, thinking out loud, cracking jokes and moving at the right pace, rather than focusing on boring shit."

In Part 1 of the series, "An Introduction to Digital Audio Production," Cosm succinctly introduces Ableton Live's GUI, explains basic sound design and audio routing, how to input MIDI events, draw automation, filtering, EQing, and part layering (Cosm, 2009). He begins with one of Live's built-in synthesizers, Analog, "because it's nice and clear—it gives a nice line of routing from where the audio starts to where it finishes . . . so you can get a really clear idea of how synthesis works." Cosm builds a sound using the oscillator section of Analog, explaining that "an oscillator is the very beginning of a sound" that takes its name from a simple shape that oscillates, "and as the shape repeats it's going to generate a tone for us." Cosm mouse-clicks in a few MIDI notes to create a bass part using a saw wave sound, and compares its sound and shape to sine, square, and sawtooth waveforms, using onscreen diagrams and Ableton's Spectrum Analyzer plug-in to illustrate the differences among them.

Cosm then explains the Filter section of the Analog synthesizer and shows how to draw automation to alter its settings as the bass part plays. Finally, he shows how to duplicate the saw wave bass sound and change the duplicated part's waveform to a sine, so that each bass part inhabits a different region of the frequency spectrum. This allocation of different parts to different frequencies is key to producing tracks with clarity. "We want to make sure that this channel is doing all of the sub [frequency] work, and this channel is doing all of the high [frequency] work," he explains. "We don't want any kind of crossover between the two." For the saw wave bass, Cosm shows how to use an EQ to remove unnecessary frequencies from its sound. In sum, Cosm's knack for explaining fundamental production principles removes their mystery. By the end of the 60-minute video he has demonstrated an ever-relevant technique for creating a rich composite sound from a few simple production moves:

> So we've got two synthesizers going to create one single bass line—we've layered them together. It's a very powerful way of achieving sounds that we want. You could layer eight sounds together if you want, to create a new sound.
>
> (Cosm, 2009)

Mr. Bill: "Create a Melody From One MIDI Note"

The Australian producer Mr. Bill (Bill Day), began posting 5- to 10-minute Ableton Live tutorials in 2010. Known today for his intricate bass music, Bill credits watching Tom Cosm's videos, reading the entire Ableton Live user manual, and his own experimentation with helping him develop his advanced production skills. Among producer-fans, Bill's tutorials are renowned for showing lightning-fast use of the QWERTY keyboard and mouse for "speed editing" (Day, 2013), creative chaining of Ableton Live's stock effects for sound design, and quality finished tracks. The titles for his tutorials describe inventive production techniques which Bill arrived at through trial and error, including "Delay Time Tricks," "Bouncing Your Master Chain," "Making Complex Drumbeats Using Arpeggiators," and "Create A Melody From One MIDI Note."

In "Create A Melody From One MIDI Note," Bill shows how to use an arpeggiator to create a melody (Day, 2010). In this video he transforms a single MIDI note into an arpeggiating melodic line by chaining together three of Live's plug-in effects—Arpeggiator device, Scale device, and Ping Pong delay—into an effects chain. In 2010, this production technique was both original and also prescient, in that it demonstrated a way of producing—rapidly mouse-clicking MIDI events into Live rather than using a controller to play the parts—that has since become common practice. In the comments, one viewer notes that, by watching Mr. Bill's videos, "Ableton has come alive with the possibilities in it . . . you've made me see the potential of this software and not just the iceberg tip that blinkered me"(Day, 2010). In several episodes of his Mr. Bill Podcast, which he started in 2019 as a platform for discussing production with other producers, Bill has explained his ambivalence about some of his early tutorials. Looking back on these videos, he remembers thinking that giving away his production secrets online would result in musicians having less respect for his music. But, on the contrary, by sharing techniques Bill solidified his reputation as both an accomplished producer and a talented teacher.

Andrew Huang: "4 Producers Flip The Same Sample ft. Dyalla, Mr. Bill, JVNA"

Canadian producer Andrew Huang maintains one of the most watched channels on YouTube, with over 2 million subscribers. Huang began posting videos in 2010, mainly clips of himself singing, rapping, and playing instruments on his own songs. Since turning his attention exclusively to music production around 2016, Huang has made videos on sound design, music theory, songwriting, and gear reviews that are informative and entertaining. Huang's most popular series is "4 Producers Flip The Same Sample," in which he and three other producers make tracks using the same audio sample. "This is really about how creative minds will take the same material to completely different places," he explains in the introduction to the series. In a 2018 video, "4 Producers Flip The Same Sample ft. Dyalla, Mr. Bill, JVNA," Huang enlists the producers Dyalla, Mr. Bill, and JVNA to work with a 20-second sample of a song by a fellow YouTuber and singer-songwriter, Malinda (Huang, 2018). The format of the video is as follows: we hear the original audio sample, each producer briefly explains the techniques they used to rework it, and then we hear their finished tracks. As we listen we watch reaction shots of the producers listening to one another's work.

In this episode, each producer has their own preferred workflows, although they all use Ableton Live. JVNA, a producer from Los Angeles, explains how she combined her own hip hop beats with part of the sample: "I duplicated it, multiplying it many times; I also pitched up the key." Dyalla, a producer from New Zealand, explains how she put the sample into Ableton's Sampler, pitched it higher as did JVNA, and "then EQ'd out the low end of the sample; this gave me more room to make up my own baseline to manipulate the song so it sounded like a happy song." Huang is next, describing his work as "a Drake-type beat" which he achieved by high-pass filtering "out all of the treble so it's just this low-end bed for the track." He also chopped small slices of the sample to use as percussion sounds, isolating "the tiniest little piece of the 's' sound and that is our hi hat." Last is Mr. Bill, who explains how he created two pad sounds from the sample in Ableton's Sampler, one pad "that's looping over four bars, and one that's looping every quarter note, which gives the tune a pace from the get go." Like Huang, Mr. Bill uses snippets of the vocals in the sample, processing them with reverb to create the timbre for a chord progression to which he applies "a few different glitching effects." Bill's beat elements are derived from a process he calls "re-grooving" the vocals bits, while his bass comes from resampling a slice of the sound of his entire mix bounced down to an audio file.[1]

What is noteworthy about the Flip A Sample series is that its episodes show how fundamental resampling (a technique discussed in Chapter 6) is to contemporary electronic music production. Each producer relies on Live's sampling instruments, Sampler and Simpler, to create pads, percussion, and other timbres. These samplers allow the producers to quickly re-pitch, filter, and loop sounds, effectively providing them with a meta-instrument made from the sample material. Even though there are stylistic differences among the producers' tracks and the video does not reveal specifics about any third-party plug-ins they use in their work (except for Mr. Bill, who mentions that he used a Roland TB-303 emulation, Audiorealism's Bass Line 3, in his track), the producers draw on a shared production vocabulary that includes sample chopping, dramatic EDM-style build-ups, breakdowns, bass drops, and generally dynamic arrangements compressed into sub-three minute song forms.

Viewer comments on Huang's series are enthusiastic about the depth of production knowledge on display that is performed through the producers' tracks, and also in the producers' reaction shots whilst listening and head nodding to the work of their peers. A viewer named Leigh Charles speaks for many producer-fans when he says, "I never knew music production was like this. . . . I just thought being a producer was just recording and mastering stuff. Not like creating the beats and taking a sample like this and creating something new." Another viewer named goldhillproductions, a music technology teacher, credits Huang's series for helping his/her students understand the potential of samples for music production-based composition. "It's a difficult learning curve approaching such a seemingly abstract concept like sample manipulation," s/he says, "and because of the infinite possibilities it's also such an individual thing which is hard to teach without diverse examples. You have created (unwittingly) a brilliant teaching resource with all that in mind" (Huang, 2018).

Deadmau5: "Deleted deadmau5 Chord Progressions"

On YouTube there are several videos of Canadian producer Joel Zimmerman (deadmau5), one of the most commercially successful EDM artists, streaming music production and DJ sets from his home studio. The videos are not tutorials per se, but rather live streams that originally appeared on twitch.tv, a streaming platform for video gamers which also showcases electronic music producers. Zimmerman's streaming sessions ranged from 30 minutes to over 9 hours in length. On the YouTube channel Oskillator, there are several videos excerpted from Zimmerman's twitch streams. In one video, "Deleted deadmau5 Chord Progressions," the producer uses Xfer Records's Serum wavetable synthesizer and Cthulhu arpeggiator plug-ins in Ableton Live to create an arpeggio on a rather enchanting chord progression (Oskillator, 2017a). As the chords arpeggiate, Zimmerman changes their sound by adjusting Serum's waveforms, filter settings, and Cthulhu's arpeggio pattern presets, listening and adjusting. As producer-fans watching him tinker with patterns and timbres, we consider what Zimmerman might be thinking as he experiments. Is this an idea for a piece, or just fooling around? Why does he like this waveform shape and not that one from a moment ago, or is he just comparing the two? How does the producer know which sound is the *right* sound?

The Oskillator channel includes a link to Zimmerman's original stream, which shows that the arpeggio experimentation portion of the session was in fact 42 minutes long (Zimmerman, 2015). Watching this extended footage, we learn new things about how the music came to be. We see how Zimmerman built up the chord progression by mouse-clicking in one MIDI note at a time, how he used sustained tones before adding the Cthulhu arpeggiator to rhythmify them, and how he tried out numerous configurations of the progression in different keys and registers (by dragging its MIDI up and down Live's MIDI Editor page) before finally deciding to delete the arpeggio and move onto another musical idea. Viewers appreciate Zimmerman's sharing his workflow. In the comments for another Oskillator video excerpted from the same stream (albeit titled differently: "deadmau5 makes THAT SYNTH SOUND . . . and junks it"), a producer-fan named unfa comments:

> Big respect to deadmau5 for having the courage to show his raw workflow. Most artists only dare to show their final tracks, to make sure they make the best possible impression. But the truth is—everything sucks in the beginning. Iterative refinement is key to making good stuff.
>
> (Oskillator, 2017b)

In sum, unfa is right: the extreme length of some of Zimmerman's original twitch streams makes them a realistic representation of how a producer's "raw workflow" unfolds in real time through iteration. Perhaps more than any of the other videos discussed in this Interlude, Zimmerman's workflow shows how in production, as in most endeavors, it is often the case that *excellence is mundane*, "accomplished through the doing of actions, ordinary in themselves, performed consistently and carefully, habitualized, compounded together, added up over time" (Chambliss, 1989: 85). Zimmerman does not talk about what he is doing, nor does he ever appear to be in a rush to finish a track. Such extended, meandering videos are a fascinating corrective to the tendency among some YouTubers to make their tutorials brief, the better to keep our attention. Zimmerman takes his time, and the educational-artistic value of his streams lies in their bearing witness to real time experimentation within a musical system, as the producer shifts among different workflow moves, trying out sounds.

Virtual Riot: "Virtual Riot Making a Song From Start to Finish!"

Christian Valentin Brunn (Virtual Riot) is an influential German electronic bass music producer who became a popular YouTube teacher through his "In The Studio" tutorials that show his advanced sound design using Serum and Ableton Live. As fellow producer Jacob Stancazk (Kill the Noise) explains Brunn's influence,"*I'm just gonna open up this synth and just show you what to do exactly*. That shouldn't be underestimated: that kid and that synth together really sculpted an entire generation of post-Skrillex dubstep [producers]" (Stancazk, 2020).

In his 2020 video, "Virtual Riot making a song from start to finish!" Brunn sketches the elements of a dubstep/trap-style beat in 50 minutes, using Ableton Live, the Serum synthesizer, and a Mac-Book Pro, monitoring his mix through the laptop's built-in speaker (ncross10, 2020). Like Mr. Bill, Brunn works fast, and his video is notable for showing (1) how fluidly he designs parts for a track and (2) his resourcefulness in using a minimum of plug-in effects to design a varied set of sounds. Using his computer's trackpad as a controller, Brunn clicks his way around his project, browsing, loading, and editing samples, makes sounds in Serum, and arranges the elements in Live's Arrangement page. Brunn rarely deliberates whether or not a sound is ideal; instead he micro-edits as he goes along. Within a day of the tutorial's release, electronic music producer-fans left comments praising Brunn's production skills: "God the speed of this man and his music theory knowledge is amazing, he is literally tuning everything by ear constantly and really killin it" (Gavin); "VR makes sound designing look easy as hell all the time" (Pahadi Savage); "I feel very dumb watching this he does so much automatically i am defs not there yet" (Gino); "Imagine getting all the things he's doing" (Alesh).

As with Zimmerman's stream, Brunn's video demonstrates a real time workflow and problem-solving. In under an hour, the producer mouse clicks MIDI parts, designs and layers sounds, uses Live's stock effects (such as Saturator and Auto Pan) and his own Ableton effects Racks, arranges the track, and through the process happens upon what he calls "happy accidents." His video offers six workflow concepts that can be applied to any style of electronic music production.

Create the MIDI part first. Brunn begins his track in Live's MIDI Editor by mouse clicking in the kick drum and snare MIDI for a trap music-style beat and an accompanying rhythmic "yoink" sound effect. "I'm gonna draw the MIDI first," he says, "and then keep that on loop while I work on the patch—that's usually how I do it." With the MIDI parts on loop, he quickly auditions snare and kick sounds from his sample library and creates the yoink effect in Serum.

Wet and dry contrast. Brunn discusses the importance of incorporating wet-dry reverb contrasts in a track. As he imports a drum break from one of his own sample packs, he applies Live's reverb to give it a sense of depth:

> We could add some room [reverb setting] . . . I like putting in the shortest possible reverb times so it's like a little reflection. . . . Also great to add stereo width to a mono sound, or kind of like a delay—a pre-echo. . . . You can time it to the speed [global tempo] of the track.
>
> (ncross10, 2020)

Brunn adjusts the reverb setting on the drum break, and explains how the contrast between reverb'd wet and un-reverb'd dry sounds in a mix creates the sensation of moving between different physical spaces:

> This is also a really nice way to play with contrast in a track: have very wet elements and then very dry elements together, especially when you're wearing headphones. That's going to sound really cool and interesting because you're constantly moving from one space to another.
>
> (ncross10, 2020)

Layering. Brunn uses several kinds of layering to create richness and a sense of depth in his tracks, a topic explored further in Chapter 6. The first way is to copy a track and change an aspect of the copied track's sound. For example, Brunn duplicates his snare part and runs it through a different reverb effects chain to create a pitched sound layer:

> Once you have a snare like this . . . we're going to duplicate the whole shebang and run it through the effects chain and then run it through 100 percent reverb but really short. And

now you can either pitch the sample before it goes into the reverb, so you have a tuned reverb tail that is in key with the track.

(ncross10, 2020)

A second kind of layering is using what Brunn refers to as *top loops*, which are high frequency percussion or drum loops such as tambourines or hi hats. Top loops are used to fill out the mix and create a sense of momentum. Brunn drags and drops pre-made top loops from his sample library into his arrangement. "I don't want to get hung up on creating a top loop every time," he says, making a pitch for his own sample pack for sale. "So I made some top loops for people like us who don't want to waste their time making top loops" (ncross10, 2020).

A third kind of layering is to double a sound with an unrelated sound playing the same or similar part. Building on the sub bass part he designed in Serum, Brunn doubles it with a sample of a voice singing one note: "I wanted to have one cool layer for this bass . . . let's take a vocal sample that sings just one note" (ncross10, 2020). Brunn finds a vocal sample from his library and drags it into the arrangement, aligning it with the sub bass part. He adjusts its pitch using Live's Frequency Shifter effect, compresses the sound using OTT, a multiband compressor, and then adds Live's Auto Pan effect to create a wobbling sound:

> I like layering buzzy basses with something melodic on top; [the layered melodic sound] is almost like a background ambiance kind of thing. . . . I want to tune this [vocal sample] with a frequency shifter to be a layer on the sub . . . add OTT, but frequency shift this thing so now it sounds weird . . . and to give it a bit of a rhythm. Put an Auto Pan on it and have it wobble at 8th notes—that's a cool way to give this sub a layer.

(ncross10, 2020)

Layers of rhythmic movement. After Brunn layers his bass sounds with a vocal sample, he returns to Serum to create additional layers of rhythmic movement. He adds Serum's white noise generator to the bass sound and links the noise's volume control to an LFO (low frequency oscillator) which he automates to create a pulsating sound. It reminds Brunn of one of his earlier tracks, "Self-Checkout:"

> We can even give this some modulation—thinking of the "Self-Checkout" tune—we could add the white noise in Serum and give this a faster LFO so there's this *meh-meh-meh* modulation in there on top of the quarter notes. So again, there's different patterns at different speeds running at the same time—that's kind of cool—keeps this interesting.

(ncross10, 2020)

Random-sounding, all over the place parts. A technique used by Brunn and many bass music producers is to incorporate random-sounding samples into their tracks. These samples are the result of sound design sessions where producers generate sounds by effects processing and resampling their own work into noisy soundscapes. These lengthy effects-processed resamples, or portions of them, are then later incorporated into tracks. Brunn drags in some random-sounding samples into his arrangement, editing their timing and tempo. The samples add color and a sense of unexpectedness to the music:

> All of these are super random-sounding, all over the place—not synced to a specific tempo or anything, so I'll just drop them in, move them around until they somewhat fit, and after I found a place where they sound cool, go in and move them around or stretch them and warp them so they are actually on the grid.

(ncross10, 2020)

Make boring sounds more interesting. If there is a single workflow "secret" underlying electronic music production, it is to continually micromanage ways of transforming conventional sounds into more unusual and compelling ones. As Brunn edits some of his random-sounding parts, he makes a series of quick adjustments using Live's Saturation and Delay effects. "Let's micromanage this real quick," he says, and within seconds the track's sound design comes alive.

Conclusion

The tutorials of Tom Cosm, Mr. Bill, Andrew Huang, deadmau5, and Virtual Riot are just drops in YouTube's ocean of evolving content about electronic music production.[2] The videos offer electronic music producers of all skill levels a plethora of ideas about techniques and workflow moves for creating sounds and, ultimately, finishing tracks. But perhaps the tutorials' most useful aspect is how they combine explanation with doing: *the producers explain what they do, why they do it, and we see them doing the work itself*, watching their cursors move through the DAW environment. As producer-fans ourselves, we appreciate, yet also assess, the practical and aesthetic value of a craft that we know from our own experience is difficult and time consuming. As we follow a producer selecting this sound not that one, altering this parameter over another, dragging and nudging a sample just to here and no further in pursuit of the perfect constellation of sounds, we draw on our own production understanding to imagine their thinking and consider their decision-making. The producer's production moves, their commentary, and their sounds give us a sense of tacit production knowledge in action, inspiring us on our own production Quests.

Production tutorial Quest

Use YouTube as a strategic educational resource to widen your electronic music production frames of reference. Seek out videos that push you to new understandings of concepts and workflows for using specific tools. Make notes on what you have learned, try out the techniques for yourself, and refine your creative algorithm.

Notes

1 As discussed in Chapter 4, resampling the sound of an entire mix is a technique Mr. Bill uses to generate material for his sound design *mud pies*.
2 The popularity of these videos has inspired similar content on subscription websites. For instance, masterclass.com features music production tutorial series by deadmau5, Armin van Buuren (Tiësto), and Timbaland.

5   Timekeeping ways

Rhythm programming

Introduction

For many electronic music producers, rhythm programming/beat-making is the foundation upon which all but the most free-form ambient musics are constructed. Beats and other kinds of steady pulsation bring vitality, flow, and kinetic life to a track by organizing, subdividing, and articulating the time of the music. A compelling rhythmic design can take many forms: a beat that sounds like real drumming, a robotic tick-tocking free of real drumming's conventions, or a skittering construction in between. This chapter offers a history of electronic rhythm programming from the 1980s to the present, unpacks the anatomy of a groove, considers five rhythm programming examples, and concludes with five general rhythmic principles.

Rhythm programming history: electronic beat-making since the 1980s

Although rudimentary electro-mechanical "rhythm boxes," such as Harry Chamberlin's Rhythmate and Wurlitzer's Side Man, had existed since the late 1940s (Brett, 2016), electronic rhythm in its modern incarnation began in the late 1970s and early 1980s, when transistor- and microchip-based drum machines such as Roland's TR-808 and Roger Linn's LinnDrum became popular production tools. Pre-DAW, the development of drum machines, along with synthesizers, samplers, sequencers, and MIDI laid the foundations for contemporary electronic music production and inspired a myriad of machine-oriented styles, from electro to new wave synth pop to techno. Drum machines' *step sequencing* functionality, which represents rhythm through a row of buttons that light up in a left-to-right flashing sequence, was particularly impactful. Step sequencing rhythms helped producers move "from thinking of music in terms of harmonic progressions and conventional song structure to thinking of music in terms of *sequences*, discrete passages of sound and time to be repeated and revised *ad infinitum*" (Hamilton, 2016). Step sequencing gave producers a new frame for thinking about the patterns that make up a song's drum and percussion parts. Veteran producer Simon Thornton explains that prior to the TR-808, "you'd be tapping out the rhythm of each drum separately—but suddenly [with the 808] you could see the beats in front of you represented by flashing lights" (Marsden, 2008). In sum, the rising popularity of drum machines was due to the fact that they functioned like a real drummer in adequate ways. Music software designer Mike Daliot observes:

> There is a reason why they were accepted as a replacement for acoustic drums. People accepted this total abstraction of drumming only because the relevant parameters behaved like they should: it's partially right, but it's right at the right spots.
>
> (Goldmann, 2015: 61)

Drum machines' knack for being right at the right spots—meaning that they could reliably keep time and faithfully reproduce or simulate drum and percussion sounds—transformed the process and sound of music production. Examples abound: as used by Afrika Bambaataa and Arthur Baker, Marvin Gaye, Prince, Trevor Horn, and Public Enemy, among other production innovators (Brett, 2016, 2020; Moody, 2012) drum machines made possible sounds and rhythmic textures otherwise unplayable by a human drummer, freeing a new figure of music production, the drum programmer, who could "avoid the tried and tested conventions that the body unthinkingly repeats" (Goodwin, 1998: 125).

A new kind of musician: the drum programmer

One of the early drum machine experts in the 1980s was a drummer and drum programmer named Jimmy Bralower, an in-demand musician who collaborated behind the scenes on recordings for Madonna, Cindy Lauper, Hall & Oates, Kurtis Blow, and others. During this time, the sound of popular music's drumming was changing due to the influence of drum machine beats. These beats were often programmed by musicians without drumming experience or an understanding of a drummer's sensibility. Bralower observed how this new way of working with electronic rhythm created "a whole other world of beats" that sounded compelling:

> When I started programming, producers and artists would come to me looking for the drum machine to sound more like a human drummer. But once hip-hop artists and DJs and people who weren't necessarily considered traditional drummers started sitting down at these machines and programming beats without a drummer's sensibility, they would come up with rhythms that just sounded cool in the machine [that] had nothing to do with playing the real drums. When that started, you had a whole other world of beats, and a whole other palette. At the same time, sampling came in. Everything we did was affected by that moment in technology.
>
> (Ampong, 2018)

Bralower, who quickly became an expert of the LinnDrum, Roger Linn's follow-up to his LM-1 drum machine, also noticed a dilemma drummers faced upon the LinnDrum's arrival: "It was like [producers] wanted the real guy to sound like a machine and they wanted the machine to sound like a drummer" (Amendola, 2010). As a compromise situated halfway between the human and machine, Bralower devised unconventional techniques for making his LinnDrum beats sound more expressive. For example, when programming drum fills, he kept the hi hat playing, something a real drummer would never do:

> When I used to do drum fills on a LinnDrum, I used to keep the hi hat playing because when you stop the hat to do a fill, it just sounded really unnatural. Somebody would say, "Well, a drummer wouldn't do that." And I'd say, "Well, if the drummer had three hands he would."
>
> (Ampong, 2018)

A similar musician-meets-machine dynamic came into play with the design of the E-mu SP-1200, a drum sampler released in 1987 whose 10-second record time and gritty 12-bit resolution would inspire hip hop producers to "craft a complete instrumental on one portable machine" (Detrick, 2007). One of these producers, Hank Shocklee, a member of the production team The Bomb Squad, used the SP-1200 as part of his musical system for making beats for Public Enemy. Shocklee noticed that the machine was configured so that its open and closed hi hat samples cut off one another when quickly triggered in succession, to mimic the sound of a drummer playing a single hi hat in open and closed positions. When Shocklee loaded other, non-percussion sounds onto the channels, these sounds also cut off one another when triggered. Shocklee paid attention to the unusual, clipped sound. Like Bralower keeping his hi hat parts playing through drum fills, Shocklee pursued the SP-1200's potentials. "Now the motion of having one sample cutting off another sample," he recalled, "creates a new rhythm, a new vibration, a new frequency" (Kelley & Muhammad, 2015).

In the 1980s and 90s, a proliferation of electronic dance musical styles—electro, techno, Miami bass, hip hop—was powered by musicians who, using electronic rhythm technologies in ways not imagined by their developers, programmed grooves that made "no attempt to reproduce the patterns that would be deployed by a real-time drummer" (Goodwin, 1998: 127). Roger Linn recalled that even though he had designed his digital LM-1 and LinnDrum to sound realistic and differentiate them from Roland's analog-based and synthetic-sounding TR-808, he was surprised to hear musicians intentionally making his machines "sound as inhuman and rigid as possible" (Bencina, 2012). The inhuman/rigid sound of electronically produced rhythms swiftly redefined drumming and created "a whole set of genres of beat-oriented music that continues to the present day" (Warner, 2017: 151). One of the most cited examples of this new rhythmic reality was New Order's "Blue Monday," a synth pop/techno song from 1981. The band programmed the song's beat on the Oberheim DMX, a digital drum machine that took after Linn's LinnDrum. (Fittingly, an advertisement for the DMX in *Keyboard* magazine described the machine as a "new breed of instrument" and "Un-Real Drums"). New Order's DMX beat had a stream-of-16th-notes kick drum pattern on beats three and four of every second bar. The band's vocalist, Bernard Sumner, would later refer to the track as an *experiment in technology*, and in fact, "Blue Monday" is exactly that: a four-on-the-floor, proto-dance music featuring a drum machine pattern idiomatically unlike anything a real drummer might (or could) play.

In his 1999 book on the intersection of black musics, science fiction, technology, and Afrofuturism, *More Brilliant Than The Sun*, Kodwo Eshun offers a crucial insight: drum machines and other forms of electronic percussion were indeed new forms of musical instruments because they never *sounded* like real drums in the first place, which is why they were used in new ways. The sound of the machines changed how musicians thought about rhythm, and the capabilities of the machines explain why Bralower had adopted the programming rationale, *if the drummer had three hands he would*. Eshun:

> There are no drums in [drum machines]. . . . You'd listen and they'd sound utterly different from drums. The movement from funk to drum machines is an extremely incredible one: people's whole rhythmic perception changed overnight.
>
> (Eshun, 1999: 186)

Charting the "movement from funk to drum machines" from the 1980s onwards, Eshun identified a shift in electronic music production whereby rhythm or beat-making as a function of the body dexterity of the drummer became rhythm as something programmed via an electronic device and, eventually, software. Moreover, rhythm's electrification and digitization brought about a change in the relationship between the musician and rhythm's patterns. In acoustic drumming, gesture and rhythm are connected, insofar as a "techno-physical, or bio-mechanical, relationship develops between the kit and the drummer's body, between the instrument and the playing techniques" (Avanti, 2013: 483). By contrast, a producer's rhythm programming mostly decouples gesture and rhythm. Over time, new ways of programming rhythm—by pressing buttons on a drum machine, finger drumming on pads to trigger drum sounds, or as is common today, mouse-clicking inside DAW software to sequence beats—largely freed physical drumming of its drumming conventions.

"Just Leave Quantize Off:" the MPC and J Dilla's groovology

Building on his success with the LM-1 and the LinnDrum, in 1988 Linn collaborated with Akai on the MPC60, a hybrid digital drum machine, sampler, and sequencer. With its 16 large rubber pads for tap-sequencing samples into songs, the instrument steered the production workflow of hip hop. An architect of this new workflow was J Dilla, one of hip hop's most innovative producers. J Dilla learned to use the MPC60 and then a later model, the MPC3000, recording his beats by finger drumming patterns on the MPC's pads. In the 1990s and 2000s, Dilla created tracks for a who's who of hip hop and R&B, including A Tribe Called Quest, De La Soul, Erykah Badu, The Roots, and The Pharcyde. His beats have widely influenced the practices of beatmakers ever since and are renowned for their warm and low-fi sound, as well as the feel of their subtle timing qualities. J Dilla's approach to his craft inspired producers to think about how certain machine-made grooves move the way they do. For instance, on the forum futureproducers.com in 2008, a producer-fan named samplesbank asks:

> So I listen to a lot of hip hop and noticed cats like j. dilla, madlib, black milk, flying lotus and a bunch of others . . . they have this off-beat sound to their tracks like the snare is late or early and the high-hats seem off but on at the same time. . . . I've been trying to get that sound by using no quantize and having the metronome off . . . but still can't get that vibe. What's the secret???
>
> (samplesbank, 2008)

Perhaps, samplesbank wondered, producers are "shifting all the snares a tiny bit early?" Another producer-fan named guilty j explains how J Dilla produces:

> [The producer is] not shifting anything he's just playin live like a real drummer would. Just leave ya quantize off and play in a good rhythm, you'll get that off-beat sound.
>
> (samplesbank, 2008)

The timing of J Dilla's "off-beat" beats is elastic with an imperfectly perfect quality, a rhythmic feel that the drummer Questlove describes as "this whole drunken-style-but-staying-on-beat thing" (Mao, 2013b). By many accounts, J Dilla did not rely on the MPC's Timing Correct *quantization* options, and only sparingly used the machine's various shuffle settings when programming his beats. "He was programming it, but it just felt live," notes the rapper Q-Tip. "The swing of it . . . the way he had the swing percentages on his beats" (Mao, 2013a). The producer-fan guilty j has it right: instead of quantizing, J Dilla mostly relied on finger drumming skills to create timing nuances. Like the feel of a skilled drummer's playing, J Dilla's beats capture a loose, yet tensioned rhythmic interlock within the drum pattern's snare, kick, and hi hat parts and other samples so that relationships among the sounds weave complexities and micro-discrepancies into the mix (cf. Keil, 1987). As heard on tracks such as "Lazer Gunne Funke," from 2009, J Dilla's beat-making feel is difficult to replicate. Kanye West memorably describes its perceptual magic: "With Dilla, every time, it's like that kick just sat so perfectly," he says. "And his swings, his shuffles on his beats, his snare choices, the way he sampled shit . . . It felt like drugs" (Sanders, 2017).

Just attack velocities: making beats with digital samplers and DAWs

With the success of the Akai MPC and other digital hardware samplers, electronic musicians in the 1990s turned to sampling for designing intricate percussion parts. Producers of drum and bass manipulated funk drumming *breakbeat* samples such as the "Amen" drum solo break from The Winstons' 1969 song, "Amen, Brother." On tracks such Roni Size & Reprazent's "Matter Of Fact," from their 1997 drum and bass recording, *New Forms*, sped up breakbeats are rearranged into stuttering, hyper-syncopated permutations. "Virtuosic in its programming" (Toop, 1999: xxx), drum and bass beats had a sound that critics at the time described as "body-baffling and discombobulating" (Reynolds, 1999: 254). In an essay on electronic rhythm, Rick Moody connects producers' transition to digital sampling with ever more abstract percussion programming. "It's the drums that really changed with digital editing," he points out (Moody, 2012: 411). Eshun explains the elements of this emerging non-acoustic-based and misinterpreted percussive abstraction producers were exploring:

> There are no drum-machines, only rhythm synthesizers programming new intensities from white noise, frequencies, waveforms, altering sampled drum sounds into unrecognizable pitches. The drum-machine has never sounded like drums because it isn't percussion: it's electronic current, synthetic percussion, syncussion. . . . Calling the rhythm synthesizer a drum-machine is yet one more example of [r]earview hearing. Every time decelerated media writes about snares, hihats, kick drums, it faithfully hears backwards. Electro ignores this vain hope of emulating drums, and instead programs rhythms from electricity, rhythmatic intensities which are unrecognizable as drums. There are no snares—just waveforms being altered. There are no bass drum drums—just attack velocities.
>
> (Eshun, 1999: 78–79)

With the improvement of computer processing speed in the late 1990s and 2000s, producers turned to using DAW software (Chapter 1), which made possible increasingly precise rhythm programming and timing manipulations in their beat-making. Among its many other powers, the DAW, note R. Brovig-Hanssen and Anne Danielsen, "presented new opportunities for optimizing and experimenting with the micro rhythmic design of grooves" (Brovig-Hanssen & Danielsen, 2016: 111). The slightly askew beats in R&B and hip hop tracks by Brandy and Snoop Dog, for example, showed the DAW being used as a "tool for generating and optimizing new, as-yet unheard rhythmic feels" (ibid.: 114). Editing beats visually on a timeline inside a DAW allowed producers to experiment with "a new complexity in the structural and micro temporal features of computer-generated groove-based music" (ibid.:114). A pivotal producer during this era, whose work synthesized hip hop, R&B, and electronic dance music, was Timothy Mosley (Timbaland). In his music for Missy Elliot, Aaliyah, and Justin Timberlake, Mosley created beats by manipulating unusual samples in the DAW, Pro Tools. Tracks such as Aaliyah's "Rock The Boat" and Timberlake's "Cry Me A River," both from 2002, have grooves made with "drum kits" comprised of samples including birds, cooing babies, and beatboxing.

The four-on-the-floor beat

In addition to making possible unusual rhythmic designs and hyper syncopated percussion, drum machines also brought a rigidity into electronic music production, most famously in the form of the *four-on-the-floor* kick drum beat, which had first appeared in disco music in the early 1970s (Shapiro, 2015). One of the first instantiations of the disco beat was in Harold Melvin & the Blue Notes' 1973 Philly Soul hit, "The Love I Lost," which featured drummer Earl Young deftly playing four-on-the-floor kick and open-closed hi hat ostinatos. Around the same time in Europe, the Italo-disco producer Giorgio Moroder became an early popularizer of four-on-the-floor. As Moroder tells the story, he didn't have a drum machine in the early part of his career, so he recorded a drummer playing kick drum along to a Wurlitzer Side Man drum machine used as a click track, for 20 minutes at a time. Moroder used this technique to record the four-on-the-floor kick on Donna Summer's 1977 song, "I Feel Love." The idea of working with what he called a "metronomic beat" was inspired by watching lounge bands play along to drum machines:

> I'd gone to a lot of clubs where these peculiar Italian bands played a schmaltzy kind of music and they used to have this little drum machine where, if you just pressed a button, it would play a samba, or if you pressed another button it would play a waltz. It was very basic and it had a horrible sound, but of course it played in time, so we sent out for one and we laid that down as a track. This then provided us with a four-minute, metronomic beat that had a kind of groove going on . . . that enabled us to stretch it to a 16-minute version, kept in perfect time.
>
> (Buskin, 2009)

Finger drumming dynamically interacting imperfections

For *Plentitudes*, my goal is not Moroder's perfect time but rather to capture as many performance subtleties as I can through finger drumming my beats. I want to avoid using quantization—snapping each note to the nearest beat subdivision—and agree with producer Sam Shepherd (Floating Points)'s assessment that "quantization ruins everything" (Beatnick, 2015). I drum in the percussion sounds, tapping the pads on my grid controller along to a click, improvising patterns, and recording their MIDI. My finger drumming is never perfectly on the click and I will edit the patterns later, not by quantizing, but by nudging individual MIDI notes forward or backwards a smidge, if needed, so that the rhythms sit better.

I could work quicker and achieve rhythmically tighter results by using drum loops, drum machine step sequences, or by quantizing. But such locked-in rhythmic textures would eliminate my performance's interesting inconsistencies. Ideally, a producer builds upon inconsistencies instead of smoothing them over, allowing them to trigger further micro timing variations in subsequent parts as they are added. In this way, a composite texture of accumulated, un-perfect parts is woven from *dynamically interacting layers of mistakes*. As with the accidents of improvisation and sound design, the inconsistencies in my finger drumming are complexities upon which to build. Jon Hopkins expresses this idea, suggesting that, in the DAW, the producer can use the "less accurate" timing quirks of drumming as a groove template by which to synchronize subsequent parts in a track:[1]

> I've drummed them [drum parts] basically, and rather than correct my performances—sometimes I even make them less accurate because it's those micro-shifts in timing that makes music sound good to me . . . I'm really interested in the intersection of machine-music and human performance. And with the software now you can really easily drum any kind of groove and match everything else to that groove, and it's a really satisfying thing to listen to I think.
>
> (Carroll, 2018)

Anatomy of a groove

(1) A Construction in time, a beat sculpture

No matter what style of music a producer makes, playing or programming a beat whose timing feels steady yet flexible is a difficult task. Yet there are principles to guide us. First and foremost, a groove is a construction in time that creates a flow with a forward-moving energy out of the interaction of the groove's elements. Any regular sound can be a groove. A metronome's click is a kind of groove, though not a very interesting one. A jazz drummer keeping time on the ride cymbal is a more interesting groove, because of the many ways drummers vary their timekeeping. The four-on-the-floor kick drum in EDM is yet another. Whatever its form, the most apparent qualities of a groove are its tempo and timing. Is it slow or fast? Is it quantized steady or does it push or drag in subtle ways? Does it sound human- or machine-like, or somewhere in between?

Producer Robert Henke (Monolake) suggests we think about beat-making in terms beyond the four-on-the-floor kick drum rhythm and other production conventions and clichés. He explains that the groove of rhythm programming need not derive from the repetition of unchanging parts, but rather from the elastic interaction of several elements to create a construction in time he calls a *beat sculpture*:

> If you look at a normal 4/4 pattern, if you look at one bass drum, and move it away from 1, 5, 9, 13, if you just move one bass drum away, the whole thing falls apart. The only thing you can do is to add off-beat elements. But it's always the idea of the whole straight programming, that you need to add additional elements to make sense out of it. As soon as you look at these double time, half time, broken grooves, you realize, or I realized for myself, there's much more freedom. Because what constitutes the groove is not the bass drum itself, it's the interaction between the bass drum and the snare.
>
> The whole groove, if you move yourself mentally away from a straight bass drum, you're looking at a groove as this kind of complex interaction of elements, and suddenly it all becomes elastic. And it's a little bit like a game of chess, you move the bass drum somewhere else, then you realize, that's cool, now I can add a clap at this part here, and it makes sense again. We're finally reaching this process of creating this beat sculpture again, you cut off a part at one part of the sculpture, then you realize it's not that the sculpture is falling apart, but you need to change something at a different part of the sculpture.
>
> (Walmsley, 2010)

(2) A limited set of sounds in tension and in dialogue

A second quality of a groove is that it comprises a limited set of sounds in tension and in dialogue with one another. In dance-oriented electronic music styles, the most common percussive elements are samples or synthesized simulations of kick drum (low pitch), snare drum (high pitch), and hi hat cymbal or shaker (even higher pitched) sounds.[2] High-pitched percussion such as hi hats or shakers often function as a steady *timeline* that provides a backdrop for a dialogue between the kick and snare or similarly timbre'd sounds. As we learned in Chapter 3, DAW software contains hundreds or thousands of drum and percussion sounds. Since it is difficult to get to know and remember so many sounds, it is common for producers to devise ways to limit their percussion palette. For example, one might choose to work with just a few trusted sounds such as sampled breakbeats, dance music's ever useful TR-808-style kicks and TR-909-style snares, or custom libraries created through sound design. Since each drum/percussion sound embodies a unique effect, even a single sound can potentially suggest a mood or direction for a track. But more important is for the producer to commit to a limited set of sounds that can "talk" to one another through their rhythmic interplay.

(3) An idea over time

A final quality of a groove is that, like a melody or a chord sequence, rhythm is an idea that can be developed over the time of the track. For the sake of efficiency, many producers simply *repeat the beat by looping it* to quickly dial in a steady and unchanging rhythm upon which to build other parts. This is common practice. But although this produces reliable (if sometimes tedious) results, there are ways by which to develop a beat so that instead of remaining a static loop, its groove grows, evolves, fractalizes, disintegrates, or otherwise transforms itself throughout a track. Whether working with a 4-bar loop or a 42-bar phrase, the producer can perpetually *alter a programmed groove in small ways so that it varies itself over time.* For example, individual notes or longer phrases can be transformed through effects processing and editing—delayed, reversed, muted, re-pitched, bit-crushed, played double or half time—to make an ordinary groove unrecognizably fascinating.

Five rhythm programming examples

As alive and organic as possible: Jon Hopkins

Some electronic music producers resist, and deliberately work around, the perfect timing options so easily achieved with electronic percussion by building grooves through a process that combines an initial spontaneous performance with extensive, after-the-fact editing. This approach is exemplified by Jon Hopkins in his ever-changing techno tracks. To make his 2013 and 2018 recordings, *Immunity* and *Singularity,* Hopkins' incorporated layers of percussion samples into parts that sound alive in the way their sonic micro details continually evolve and shape-shift over time.

Hopkins uses groove-making to access ways of using his musical system, "to manipulate the technology to make it sound alive and as organic as possible, and not get stuck in the world of grids and over-programming" (Future Music, 2013). On his 2013 track "Open Eye Signal," for example, Hopkins incorporates numerous subtle, yet audible complexities into its beat, such as a shifting snare drum part that sometimes hits just before beats 2 and 4, and sometimes on them, its pattern always unpredictable. It "sounds like quite a simple groove," Hopkins explains, "but there's a lot of nuances in there trying to make it sound alive, living and breathing—like it's a thing that's growing in itself" (ibid.).

Alternate meters: Jlin

Some producers combine classic electronic drum sounds with meters outside of dance music's default 4/4 time. On her 2017 recording *Black Origami*, the producer Jerrilynn Patton (Jlin) crafts rhythmically kinetic bass music tracks influenced by footwork, a fast dance and music style with syncopated double-time feels, triplet-feel polyrhythms, and sub bass lines. *Black Origami* was noted for the rhythmic complexities of its "volatile beat patterns and otherworldly fragmented sounds" (Gaillot, 2017), and the recording is in fact a kind of drum programming étude, in that all of its tracks are built from a limited palette of percussive sounds. A key to Patton's beats sounding tensioned yet ambiguous is that 11 of the 12 tracks on *Origami* have a 6/8 rather than 4/4 metrical feel. On the track "Enigma," for example, the producer sequences staccato samples of drum hits, claps, finger cymbals, and vocal chops to create pointillistic 120bpm groove textures in a six-beat meter. Like the timeline and interlocking rhythms in West African dance drumming, the beat groupings of *Origami*'s textures can be felt from multiple perspectives (e.g. two groups of three beats, or three groups of two). Patton uses polyrhythms to make the music perceptually dramatic, yet elusive.

Rhythmic echoes: Nils Frahm

The producer Nils Frahm makes keyboard-based music using a changing collection of vintage analog electronic music hardware, including the Roland Juno-106 synthesizer and Space

Echo delay. Like Patton, Frahm often works in meters with a 6/8 feel. Additionally, instead of using conventional percussion sound sets of electronic music (e.g. kick, snare, hi hat), he builds rhythms out of pulsing melo-harmonic or noise textures. On Frahm's 2018 recording, *All Melody*, several tracks are built upon a synthesizer arpeggio, a bass line, or organ stabs that cycle around in a six-beat meter. These parts are multiplied by echo effects that send pitches bouncing into polyrhythms, evoking the ricocheting double-time snare hits of drum and bass. Frahm's music moves through melodic and harmonic changes, but also shift rhythmically through various slow-changing filters opened and closed to alter timbres. For example, the tracks "A Place" and "#2" do not use drums or drum sounds, yet the music's rhythmic feel is continually accented and syncopated. Frahm's dub-like use of filter effects and lack of a four-on-the-floor beat keeps the music pulsating and open-ended.

Slippery, ghostly beats: Burial

Emerging in London in the early 2000s, dubstep relocated the shuffling/jittery breakbeat rhythms of two-step garage into a nostalgic yet futuristic sound in which "percussion was now peripheral while the bass established the music's heartbeat" (Walmsley, 2009: 87–88). An influential producer working in this context was William Bevan (Burial), who released his acclaimed self-titled album in 2006 and its follow-up, *Untrue*, in 2007. Bevan's evocative recordings "conjure an atmosphere of eerie space" (Reynolds, 2017), as if distant echoes of earlier electronic music styles, including dub reggae, jungle, and grime. To produce his tracks, Bevan eschewed conventional DAW software, using instead the audio editing program Sound Forge to sequence his beats. In Sound Forge he programs rhythms out of samples laden with noise and hiss to create atmospheric textures. Derek Walmsley identifies the sound and timing of Bevan's beats as keys to his music's perceptual magic:

> his percussion patterns are intuitively arranged on the screen rather than rigidly quantized, creating minute hesitations and slippages in the rhythm. His snares and hi-hats are covered in fuzz and phaser, like cobwebs on forgotten instruments, and the mix is rough and ready rather than endlessly polished.
>
> (Walmsley, 2009: 92)

The textures and slippages of Bevan's beats can heard on "Archangel," a track whose drum sequences consist of muted kick, noise bits, and cross stick samples arranged to sound like disintegrated breakbeats. Interestingly, Bevan thinks of his drum programming as *supporting* the other samples in his music: "The drums are more about trying to thread sounds and vocals together, they flicker across the surface of the tune" (Fisher, 2012).

Balancing uniform and variant: Objekt

In a Q&A on an online dubstep forum in 2012, producer TJ Hertz (Objekt) explained the principles of his production process, including rhythm programming.[3] Under the topic "On 'realism,'" Hertz broaches the perennial production problem of how to "add a human touch" to programmed rhythms to make them sound organic and to add "sexiness" to otherwise mechanical grooves. He recommends seeking a balance between consistency and variation by distilling those elements that make a live drummer exciting:

> Often I think finding the right balance between uniform and variant is not always about trying to emulate the live drummer exactly, but rather distilling certain aspects

of his/her performance. To me, the most interesting thing is trying to find the sexiness within such a mechanical groove, like writing music for robots to dance to, rather than trying to make a drum machine sound like a drummer. Techno is inherently about repetition, and much of the power comes from the mechanical nature of the programming.

(Hertz, 2016)

Hertz cautions, however, against trying to make rhythm programming sound like a live drumming performance by over-tinkering with its parameters, such as volume (velocity):

In my opinion you need to be careful when trying to emulate the rhythmic characteristics of live instruments within a genre like techno, especially with regards to percussion, since it doesn't always work that well. Using a super high-quality velocity-triggered randomized patch for bongo and conga loops is one thing, since bongo and conga loops are usually played by hand anyway, but I've very rarely used more than one velocity layer for a kick drum. Accenting (using more than one velocity level) is tremendously important to the groove, but a live drummer will play every hit at a different velocity and such variation in level can sometimes be detrimental to the energy of a rather mechanical loop. My compromise is often to accent by extra layers (different samples) on certain notes rather than simply increasing the velocity, or to limit the velocity levels to two (soft and hard) rather than a continuous spectrum of 0–127.

(Hertz, 2016)

To illustrate, on the 2018 track "Silica" Hertz distills his own rhythm programming advice for making mechanical grooves sound organic by going one step further that most producers. The track conveys an abstracted sense of pulse and a steady dance music tempo (123 bpm), but without ever articulating it in any of its parts. A lesson to take from this track is that sometimes the feel of a rhythm lives in its silences.

Five rhythmic principles

Play grooves

The simplest way to program a rhythm is to play it and record it one part at a time, or all of its parts at once. Play the pads of a Grid controller, the keys of a keyboard, or record your own live drumming. Boards Of Canada member Michael Sandison explains how the group records drumming to help inject "chaos" into the music:

As for our percussion, it's never just a drum machine or a sample, we put a lot of real live drumming or percussion in there, woven into the rhythm tracks, and it brings a bit of chaos into the sound that you just can't achieve any other way.

(Pattison, 2013)

Recording a groove that you play yourself has several attributes that connect to those of improvising generally, a topic we explored in Chapter 3. First, our ideas often materialize only after we have been playing around for a while—trying out patterns, experimenting, and making mistakes in search of new ways of adding to the music. *Playing sounds makes our ideas physical and helps us find combinations of sounds and rhythms that work together.* Second, playing a groove captures all of our timing nuances and imperfections, both intended and not. As we record to a click or freestyle to previously recorded parts à la J Dilla, our timing fluctuates in little ways; our concentration to stay "in the pocket" (as drummers say) leads us to further nuances. A third benefit of playing a groove is that we get into a flow, maintaining a consistent intensity while introducing variations along the way.

Layer rhythmic parts

A second lesson of programming rhythms is to layer them. Begin by recording sounds from one kit (or collection of drum/percussive sounds), then repeat the process with sounds from a second kit, maybe even a third. Layering rhythms creates complexity by giving the producer options for having sounds to work with and respond to. It often happens that layering second and third parts over an initial rhythm creates a greater composite groove. Experiment. Parts that sound extraneous can always be muted or deleted later.

Go beyond drum fills and short-term phrasing

Programming rhythms that articulate a track's pulse but not its metrical structure is a way to generate interesting musical results. Using this technique, the producer goes beyond drum fills and other forms of short-term phrasing conventionally used—especially in electronic dance music—to mark the ends of sections. Instead of fills and short-term phrasing, try programming ever-changing rhythms that interact in unusual ways. To return to TJ Hertz who humanizes loop-based and mechanical-sounding rhythm programming by configuring different loop lengths for each rhythmic element:

> Depending on the style you're going for, different elements will loop at different points. A lot of "loop techno" sounds like it loops every 2 beats but if you listen more closely the hi hats might be looping at 4 beats, there might be an open hat every 8 beats, an incidental every 16 bars, etc. It's structural details like this that can add a human touch to a very rigid framework—the stuff that makes exceptionally loopy music sound very organic.
>
> (Hertz, 2016)

Sam Shepherd (Floating Points) uses the same technique:
> The simplest but maybe most effective thing is not having the same length for every rhythm. If you've got four percussive elements, make sure they're not all the same length.
>
> (Smith, 2019)

Think in timelines and pulsation

One of the many beauties of electronic music production is that grooves can function by unconventional means. It can be useful to eschew typical groove markers, of which, besides drum fills and short-term phrasings, there are many. These markers include accentuating backbeats (e.g. in

4/4 time, the beats 2 and 4), unchanging four-on-the-floor kick drum patterns, and, returning to Henke's idea of beat sculptures, any percussive sound that robotically repeats a single pattern over and over. Devising ways to avoid relying on such markers can be a productive constraint. The challenge for any producer is how to program a groove that keeps time and a sense of momentum, yet is always evolving.

A groove need not be rigid and unchanging or ploddingly obvious with how it breaks the time of the music into smaller units. Building on Hertz's and Shepherd's using rhythms of different lengths, the producer can sidestep short looping patterns that repeat (e.g. four beats long, or one measure of 4/4 time) by instead configuring a track's rhythm in terms of *timelines* and *pulsation*. The timeline principle comes from West African drumming traditions, in which an iron bell rhythm anchors a group of drummers (Figure 5.1). The bell rhythm demarcates where one is in the cycle of time, "where a meter is suggested and played around yet unsounded" (Brett, 2017: 83). For a piece in a 12-pulse meter (such as *Atsiabekor*, one of the oldest dances of the Ewe-speaking people in southeastern Ghana, Benin, and Togo), a common timeline pattern is for the bell strokes to sound on pulses 1, 3, 5, 6, 8, 10, and 12.

The bell timeline part is not the *beat* for a piece, but rather a guiding rhythm and metrical framework within which the drummers position their rhythms (Chernoff, 1979: 48–50). The drummers' parts interweave with both one another and the time cycle of the music as articulated by the bell. Just as drummers listen to the bell timeline and fit their parts to it, the producer can devise inventive ways of using percussive sounds to move a track through timelines and pulsation, as opposed to rigidly repeating beats.

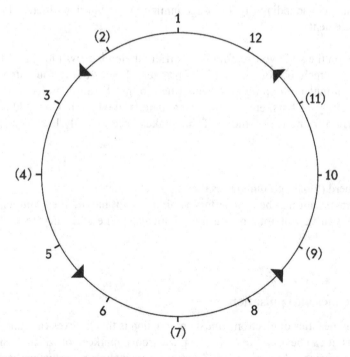

Figure 5.1 12-pulse West African timeline pattern

Edit and manipulate sounds and rhythms

A third approach to programming is to edit and manipulate drum sounds and rhythms. The sound of a beat usually remains the same over the course of a track—beginning and ending with the same snare drum sound, for instance. Historically, the reason for this is that, in rock production, the drummer played a single drum kit; or in electronic music production, the drum programmer used a single drum machine or sampler. But in the DAW era, the electronic music producer may question such constraints on sound sets. Because, following Eshun, if there are no real drums but only *attack velocities*, do one's percussive sounds need to remain fixed throughout a track? Perhaps not always. One might adjust the pitch, timbre, and envelope of every percussive sound, every beat, every bar, or every section. Or one can automate effects on drum parts so that their presence is felt in gradual and non-obvious ways. In sum, by thinking of beats as timelines and pulsations whose timbres evolve over time, the producer redefines how rhythm is manifest in the music.

Conclusion

Presciently describing why electronic rhythm was so attractive to electro musicians in the early 1980s, David Toop notes that it was the drum machine "whose artificiality liberated Electroids from drum cliches" (Toop, 2011). This theme of rhythmical-technological liberation was echoed a few years later by Andrew Goodwin in "Drumming and Memory," an essay in which he observed how electronic rhythm is free from the conventions of acoustic drumming. "If one wants not a simple groove, but something shocking and different," he said, "it is often the computer or the drum machine that can provide it, unhindered as it is by the 'thought communities' of flesh-and-blood musicians" (Goodwin, 1998: 126). Then Goodwin boldly reverse-engineered the human drummer *from* the drum machine, claiming that "the human body is simply a precursor to the drum machine—a device that remembers rhythms" (ibid.). From Goodwin's perspective, the drum machine was the natural evolution of the drummer.

Whether or not drumming has "evolved" from humans to machines, a lesson from the history of electronic percussion and drum programming is that drumming can at least be re-defined. From early rhythm machines like the Rhythmate and the Side Man to the Roland TR-808 and the Akai MPC to today's digital tools, electronic rhythm technologies have taken the physical labor out of drumming, allowing it to be programmed. The technologies have also inspired unusual ideas for beat patterns "that make references to, but do not fit within, specific pre-existing dance music genres," thereby extending producers' musical capabilities (Neill, 2004). Through their varied practices for making beats that sound realistic, synthetic, or somewhere in between, electronic music producers continue to evolve ways of programming imaginative and compelling rhythms.

Rhythm programming Quest

Make an entire track using only drum and percussion sounds. Play or program grooves, layer parts, avoid drum fills and short-term phrasing, incorporate a timeline pulse, and edit the sounds so that the rhythms go on a journey.

Notes

1 In Ableton Live, such groove templates are called Grooves and can be applied to audio and MIDI clips. Grooves can be saved to a Groove Pool list "and offer a variety of parameters that can be modified in real time to adjust the behavior of any clips that are using them" (Ableton, 2018: 197).

2 Similar pitched sound sets are in play in the rhythmic traditions of other music cultures. For example, in North Indian classical music the *tabla* drummer's right hand plays finger strokes on a high-pitched *dayan* drum, while the left hand plays the lower-pitched *bayan*. Similarly, in West African Mande dance drumming, *djembe* drummers are joined by the lower-pitched *djun-djun* drummers who provide bass tone rhythmic counterpoint.

3 The principles are compiled in a remarkable manual, "Production w/ Objekt v0.1" available via Reddit. See Hertz (2016).

6 Folding back in

Production disruptions

Introduction: the idea of disruption

Over the journey of building a track, the electronic music producer inevitably arrives at turning points in the process where the music-in-progress might benefit from being disrupted in some way in order get unstuck from its own tendencies and re-energize the producer's interest in what its sounds are doing. At such junctures in the production process, our task is to heed this call from the music—as Brian Eno puts it, having "the attention to see what's needed before it is actually needed"—by devising alternate ways to bring the sounds somewhere different from their current state (Baccigaluppi & Crane, 2011). Depending on the track and what s/he notices, the producer might need to subvert a workflow, question methods, deconstruct or further process sounds that are too ordinary or otherwise predictable, re-pitch or re-organize the sounds into a new arrangement, or make some other form of alteration. Producers often speak of the importance of changing up their workflows and sounds to keep the music fresh. Jon Hopkins describes this process of refining and tweaking sounds so they do not sound predictable as *disruption*:

> I like to kind of disrupt my own process quite a lot to make sure that things don't end up sounding predictable or, like things I've done before. . . . I will tend to just try and disrupt it in some way. That could be tuning it down, fixing it weirdly or trying a crazy new chain of plugins on it.
>
> (Rancic, 2018)

Calum MacRae (Lanark Artefax) echoes Hopkins' idea of eschewing the predictable-sounding. MacRae's motivation is simply finding ways to avoid being bored by his own work:

> It's not that I don't like making mixable or structured music, but rather when I try to do it, I find myself quite bored by what I'm doing. I don't like to listen to music that is predictably going in a specific direction.
>
> (Eede, 2017)

Similarly, Benjamin Wynn (Deru) finds that manipulating a sound often leads him down an unpredictable sonic path he could not have anticipated beforehand. He allows the results of that sonic manipulation to direct his work:

> Sometimes I'll start manipulating a sound and figure out where it wants me to go based off of the results. It can lead me down a path, and I might end up somewhere very different from where I started. You can't always predict the outcome.
>
> (Raihani, 2017)

The notion of cultivating disruption in search of unpredictable outcomes illustrates electronic music producers' continual pursuit of nonlinear results in their work. Even as they rely on their tried and true methods, producers are ever alert for ways *out of* those methods via production moves that disrupt the music and thereby make it more compelling.[1] We can think about cultivating disruption, then, as a way of re-introducing complexity into our workflows to keep the music interesting as its sounds take us on a perceptual journey whose destination we cannot predict. This chapter examines ways of disrupting the music production process that change the direction of a track via linear and nonlinear ways of working, creating variations on a single sound, folding errors back in, muting parts and isolating build ups, resampling, applying effects in layers, morphing resonances, and adding distortions or other sonic artifacts.

On linear and cyclical ways forward

In even the most dance-oriented or beat-heavy idioms of electronic music, producers grapple with finding the right mix of linear and cyclical ways of generating material for tracks (Figure 6.1). Producing linearly entails organizing the music from its beginning to its end in a way that relies on an ever-changing musical line rather than the repetition of shorter rhythmic cells. This linear way of organizing music is an inheritance from the tradition of European classical music. It generates its sense of forward-moving energy and "repetition with a difference" from the sense of "accumulation and growth" in, for example, an unfolding melody or harmonic progression (Snead, 1984: 67). The approach is encapsulated in J.S. Bach's four-part chorales, hymn-setting music from the Baroque era that is said to "progress" or create the sensation that it's *going somewhere* through chords that change one beat at a time, towards a goal in the form of a harmonic cadence.

Chorales, of course, are not particularly germane to producing electronic music. A more common approach to production is a cyclical-oriented *loop composing*, which entails organizing the music as an endless "changing same" that works by "circulation and flow," relying on repetition of short rhythmic cells more than ever-changing musical lines or harmonies (Snead, 1984: ibid.). This way of organizing music is an inheritance from the EDM continuum, whose history connects many contemporary production styles to disco and, more broadly, to African and African diasporic traditions, such as Latin American and African American popular musics, as well as jazz. These musics are often organized around cyclical repeating patterns, anchored by a *timeline* rhythm played on a bell (in African dance drumming, as discussed in Chapter 5), a clave (in Latin salsa), or a ride cymbal and hi hat (in American jazz, rock, and pop). Musics constructed upon a cyclical, changing same often have compelling, hypnotic, and funky grooves. From four-on-the-floor EDM to more rhythmically ambiguous, yet still

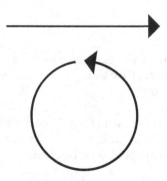

Figure 6.1 Linear- versus cyclical-based music

pulsation-based forms, a groove-oriented approach is the foundation of much electronic music production.

It is common for producers to devise ways of combining linear and cyclical musical aesthetics into their work, and both approaches have their virtues. As we saw in Chapter 5, playing or programming a beat through from its beginning to its end is one way for a producer to think in terms of phrasing and form (and also a great way to generate loop material). This approach is linear in the sense of its being sequential: our performance ushers us from one idea to the next, in an unfolding flow. Taking a linear approach is a way to develop a continuous musical line, even if that line ultimately turns out to be a composite of different sounds in a track. The producer Sonny Moore (Skrillex) describes the importance of having continuous lines in his tracks: "Even if it's a whole line of bits and crazy parts, it still has to stick with you and have . . . almost like a phrase" (Future Music, 2012a).

On the other hand, taking a cyclical approach to producing—working with repeating clips in Ableton Live's Session view, for example—can be a way to fascinating nonlinear juxtapositions. Using this approach, a producer can assemble loops of varying lengths that repeat in unpredictable and overlapping ways (recalling the generative musical systems of Brian Eno) to create kaleidoscopic counterpoint. A middle ground between the linear and cyclical approaches to production is to *use repeating parts whose timbres shift in lieu of melodic or harmonic development*. This creates a changing same in continuous, subtle flux. An enduring example of this is the filtered bass sound in classic acid house music, first used in the 1980s by Chicago producers Phuture on their 1987 track, "Acid Tracks." Phuture created this widely replicated "squelching, resonant, and liquid" timbre using a Roland TR-303 bass sequencer's saw tooth and square waves fed through a low-pass filter and envelope generator (Hamill, 2014). By incorporating ever-changing timbres via such *filter sweeps*, producers highlight variations inherent in repetition itself.

Turning the horizontal/linear into the looped/cyclical

Working in the DAW, we record a part—a beat, a chord sequence, a bass line—and it appears on the screen as a colored sequence that unfolds like an unspooled horizontal ribbon. A track with ten parts is displayed as a stack of ten ribbons, each of which contains the MIDI or audio waveforms of individual tracks. This left-to-right visual representation of the music as a sequence provides a global view of a set of linear performances and, if the producer works on a track one section at a time, these sections can be delineated, copied, and moved around in many ways.

But sometimes the linear prevents us from seeing the cyclical. Midway through producing the tracks for *Plentitudes*, I have barely explored the possibilities of my linearly composed parts arranged into horizontal sequences. There are probably many as yet undiscovered inherent rhythms I might find if I loop segments of the music (a beat? four bars? ten bars?) anywhere along the length of its unfolding parts.[2] I have resisted using looping as a form of disruption, but looping can be an escape hatch from the constraints of the DAW's left-to-right timeline environment. In a series of incisive articles about production practice, producer Mark Fell points out the shortcomings of the DAW's timeline:

> This system structures how you make music in a very specific sort of way, and it's so present that it is almost invisible. . . . I think the problem is (and was) how those timeline systems structure time—how they position you as a composer outside the music.
>
> (Fell, 2016)

Beyond the very basic recording of musical events and parameter values, the timeline producer is compelled to stop, record, play, rewind, edit, move forward. These behaviors that are quite unlike the process involved in using drum machines and other musical instruments.

(Fell, 2018)

I loop a few segments along the ribbon sequences and hear details that have been hidden within them. Looping helps me focus narrowly on a tiny slice of the track-in-progress—a slice which might reveal inherent rhythms I could develop further. To make their recording *Re: ECM*, remixes based on releases from the catalog of ECM Records, producers Ricardo Villalobos and Max Loderbauer combed through the label's recordings, looking for instrument sounds, voices, and resonances to sample. They constructed new tracks by combining their found samples with electronic sounds and processing the result, turning the linear into the cyclical. Villalobos explains how he and Loderbauer made loops out of the samples and then *improvised with the loop material*. *Re: ECM*, he says, "is an improvisation, where the elements are looped in a definite length and always repeat themselves in a certain way and also intertwine in a certain way" (Villalobos & Loderbauer, 2014).

In every track there are moments in which to explore the repetition potential of a segment of the music's horizontal unfolding. I move the loop brace around the arrangement, looping a few beats or a few bars here and there until I hear something compelling. *A great loop sounds inevitable, conveying an energy unnoticed before it was repeated.* Sometimes, the most interesting loops are of unusual lengths—say, 3.75 beats—so I zoom in on the track's arrangement to adjust the loop brace length to various off-beat locations inside the 4/4 grid, listening to the results. Any section of an arrangement can be looped as long as you find the perfect loop points. Once I locate a loop-able section, I begin muting other parts within the loop brace's span (Figure 6.2). Experimenting with different combinations of muted and unmuted parts disrupts the texture of the arrangement and reveals sound layer combinations that I had not heard until now.

Finding a good loop changes my perspective on the track I had been shaping. Suddenly there is space in the music. Suddenly I wonder about other ways the music could proceed or stay right where it is. For producer James Parker (Logos), listening to loops is a fundamental part of his workflow for making drum and bass. "I spend a lot of time listening to loops," he says. "I can sit around listening to a loop for two hours and tweaking it" (Wilson, 2019). Composer William Basinski, who builds tracks using eroded-sounding tape loops, describes the perceptual shift a good loop can create:

> Sometimes a kind of eternal perfection happens, and you can't tell the beginning and the end. It seems to create a timeless amniotic bubble that you can float in.

(Earp, 2018)

Playing with linearly-composed material through loop experimentation is a starting point for production disruption. Prior to looping bits of the track, I had been reluctant to rely on repetition to substitute for new ideas. But each loop is a new idea, a potential through-line to something exponentially more interesting, a revealing of facets of the music that were there the whole time. By looping segments in a track, the producer spins new material to play with in the Quest for hidden significances, inherent rhythms, and new perceptions.

Along with the marimbas, the lead sound on my tracks is a bell part, which sounds like a sine tone mixed with a glockenspiel and white noise. I loop four bars of the music, and solo the bell so I can audition various VST reverbs and reverb settings, turning virtual knobs this way and that to hear what happens. Some turning leads me to a crumbling sound that catches my ear. The

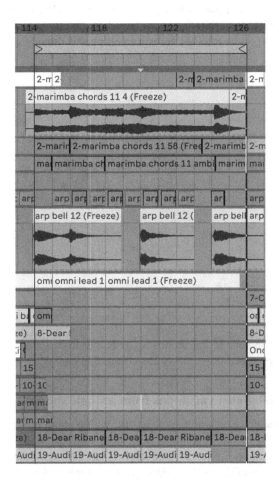

Figure 6.2 Ableton Live's loop brace with some parts muted

sound mixes a pleasing interplay between the bell tone and the reverb's eroding effects, making it unclear where the tone ends and the effects begin. I like the crumbling, reverb-ed bell sound so much that I consider deleting all of the other parts in the track and building something new with just this sound. For producer Calum MacRae (Lanark Artefax), a sound's presence always connects to its disintegration: "It seems to me that the idea of 'presence' comes about as a result of disintegration and the fracturing of a thing, it's being in between the gaps" (Inverted Audio, 2016). My response to the bell sound is a typical production moment: while doing one thing (listening to reverbs), I inadvertently came across an until-now-hidden point of entry onto the music's field of action.[3] This is but one of many examples of how electronic music production is a nonlinear and unpredictable process that offers the producer not one but many options for developing the music through disruption.

Even as I use the DAW for conventional recording and sequencing, the software's sound-shaping functionalities run towards the infinite (CPU-depending). In fact, recording and sequencing have turned out to be only a small percentage of my workflow. From this point forward, most of my time is spent figuring out ways to disrupt what I have into more compelling forms: to turn a clean bell into a disintegrating one, or make a flat sound more dimensional. In moments of disruption, the producer shifts from thinking about the music in a linear way to thinking about it in a nonlinear way. Consider some relevant verbs for putting this thought shift into action:

adjust,
tweak,
morph,
clone,
bend,
warp,
break,
mute,
double,
flip,
echo,
amplify,
stretch,
reduce,
thin,
thicken

There is much that can be done to the sounds to take them from sounding *like this* to sounding *like that*, to turn linear predictability into nonlinear unpredictability.

Spinning new constellations: variations from a single sound

It is always helpful to adopt a recombinant way of thinking about our tracks in progress, since even the most minimalist of tracks contains a plethora of information—melodies and chords, beats, samples, and atmospheric effects—that can be spun into new constellations as we shuffle the elements into variations of themselves.[4] One recombinant way forward is to *derive variations from a single sound*. Some producers use one part to play another, so that MIDI events for, say, a bass sound trigger a lead sound, or vice versa. Producer Alan Myson (Ital Tek) explains his process for using a single MIDI sequence to trigger multiple sounds, spinning timbral variations from one source:

> I find Ableton very quick for layering stuff, so I just duplicate the channels. I'll have a melody part playing out some synths that I've recorded. I'll just duplicate out three versions of it so instantly you've got a bigger sound there. . . . You can add slight variations to the second and third duplications. So often I'll reverse one of them, or I'll stick loads and loads of bit crushing on one of them just to add some texture. So you've got the same thing playing, but beefing it up so you've got more depth to the sound.
>
> (Myson, 2011)

Other producers use similar duplicative processes. Austin Collins (Au5), who makes in-depth track breakdown tutorials on YouTube, creates multiple variations of bass sounds for his tracks:

> What I usually like to do is I'll take one main sound, and I'll duplicate it and tweak that and change it into slightly different sound. And then duplicate that, tweak it, and change it into a slightly different sound. And that way everything kind of has the fullness and it sounds

like it's from the same palette. It's not just like random, completely outlandish contrasting sounds. It's like [the sounds] all come from the same family. It's a lot easier to create a bunch of sounds with diversity from a single sound and have it sound good, than create a bunch of different sounds completely independently—built completely in different ways from one another—and then trying to make those sounds work together cohesively.

(Multiplier, 2017)

Producer John Hislop (Culprate) uses the same copy-and-tweak-a-single-sound technique as Collins:

That's for me, the ideal way to build layers with synths: because if you open up a new synth for each layer, you run the risk of your layers becoming detached. And I think that's one of the most common problems with layering that people have: your layers just don't gel. So this is a really good way to get them to gel together: copying an instance you're working on and just tweaking it slightly. . . . Layering is for filling gaps in your original sound, not to change it dramatically. So I'll generally have a main sound . . . and the rest of it is just adding bits of grit here and there.

(Hislop, 2017)

Spinning the music into new constellations by duplicating sounds and altering those copies is a way to explore the latent potentials of a track by bouncing what is already happening in other directions. This allows the producer to develop sounds by recycling what s/he has by using one sound to generate another—in essence, *making mutations from a shared genetic source*. As we recycle the music it becomes increasingly fractal-like in that its elements and form are repeated and echoed in the altered copies on various levels of resolution. From samples to sequences, *any* detail of the music—down to the grain of its sounds—can be the basis for new variations. As I look at the 20 parts and ponder the many discrete MIDI and audio moments on the screen in front of me, the still unexplored possibilities for action seem endless. They *are* endless, so a useful production challenge, as producer Amon Tobin describes it, is giving oneself "a very narrow set of options and try to escape from that little box you're in and make something happen. That's when you become creative, so limitations are a great obstacle to have (Future Music, 2019).

In sum, the more the producer considers the unfinished state of the parts, the sounds, and the arrangement, the more music production reveals itself as *a practice of inventively recycling what is already present*.

Folding errors back in

Any track in progress presents the producer with opportunities for folding back in, for recycling one musical bit to use elsewhere. Even a single mistake can be folded back in.

Here and there when I am finger drumming drum and percussion parts (Chapter 5), I'll tap a pad on my MIDI controller I didn't intend to tap. I ignore the mistake and continue playing so as to not interrupt the flow. But as I listen back to the percussion parts, I notice that the errant drum hit sounds good. Although I hadn't planned for it, the mistake has productively disrupted how I hear the music. It brings to mind one of Brian Eno's Oblique Strategies: "Honor thy mistake as hidden intention."

I decide that the errant drum hit could happen more often—not every measure, but every once in a while. I copy the MIDI to various places in the part so that it can interrupt the music. From playing parts, shaping dynamics and effects, and now, making mistakes, every action I have taken in the course of building the track thus far can be folded back in. *Fold what you already have back into the music as a modulated mistake.* The production process folds and re-folds the material back upon itself, sometimes as mistakes that sound good somewhere else.

Muting kick drums and isolating build-ups

A third way to create interruption in electronic music production is to rethink a track's sonic and structural conventions, such as the four-on-the-floor kick drum and breakdown sections. On his 2018 and 2019 recordings, *Debiasing* and *Utility*, producer Sam Barker (Barker) built rhythmic tracks without the use of percussion or a steady kick drum. In considering the tendency of music producers "to rely too much on one methodology to solve problems," Barker developed a percussion-less electronic dance music sound as a re-thinking of the "the techno formula" (Kirn, 2019) and taken for granted rhythmic cues: "I wanted to attempt making dance music while avoiding the deeply ingrained cues . . . like off-beat hi-hats or 4×4 kick drums, pulling things out and putting them back in again" (Ralston, 2019). For Barker, the four-on-the-floor beat is electronic dance music's most dominant voice. It's also overused. "If you think of [the four-on-the-floor] as a voice within the realm of music-making, it's always the loudest, and it dominates over everything else" he observes. "It's also the least interesting because it's making the same statement over and over" (Ralston, 2019).

As an alternative to the four-on-the-floor kick, Barker devised a production workflow for creating a wall of sound-like harmonic texture. Then he processed that texture to make it rhythmic:

> The approach with *Debiasing* was to make huge layered drone chords with multiple poly-synths and open filters, squeeze as much energy into a chord pattern as possible, then make it spikey and percussive by dynamically processing it through the modular [synthesizer]. . . . That was the basic technique: making something as huge as possible and then reducing it down.
>
> (Ralston, 2019)

Similar to Barker's sound is the music of producer Lorenzo Senni. On his 2012 and 2014 recordings, *Quantum Jelly* and *Superimpositions*, Senni built tracks out of the arpeggiating *build-up* sections commonly found in trance, electronic dance music that features climactic *bass drops* where the four-on-the-floor kick and bass parts return or "drop" after a dramatic build-up. Senni's approach was to compose trance music-style build-ups and string them together to make a ten-sioned, rhythmic sound that builds for a drop that never arrives. Senni explains his interest in repurposing the build-up: "The build-up is interesting . . . It's functional but there's freedom in how to reach it . . . I played a couple of shows putting together these build-ups, one to the other. One would loop maybe 10 times and then the other, like a display" (Twells, 2015).

In sum, Barker and Senni's experiments with muting the four-on-the-floor kick drum and isolating/looping build-up sections while maintaining a kinetic sense of pulse illustrate inventive ways of disrupting rhythmic features of electronic music production.

Resampling

Resample constantly. Layer, process, then resample again and layer

—TJ Hertz (Objekt)[5]

The most widely used way of disrupting a track is to resample its sounds. Resampling is the practice of re-recording sounds by soloing the track(s) from which you want to re-record, and then recording its output by routing it to the input of another audio track. Once the producer has resampled a part onto a new track, the resampled part can be processed to further re-shape its sound. For example, one can make loops from it, lower or raise its pitch, add effects, or do all of these things. Resampled sounds thus become springboards for something entirely new or input for subsequent rounds of resampling. This explains the workflow used by TJ Hertz: *layer, process, then resample again and layer*. It is not unusual for producers to resample a resample several times to hear where additional layers of recording and processing might take them. In a Q&A on Dubstep Forum in 2012, Hertz shared his workflow for creating ambiances by resampling a sound several times:

> Ambiences are most often long reverb tails, resampled, time-stretched, put through more processing (heavy compression, EQ, delay, bit-crushing, more EQ, more delay and reverb, etc.), resampled again, automated.
>
> (Hertz, 2012)

The term *resampling* was originally one of several terms (other terms included *unsampling* and *downsampling*) to describe the process of *sample-rate conversion*, which changes the sample rate of a signal from one representation to another. In the early years of digital audio, the process was necessary to make different digital audio systems compatible. For example, Compact Discs and Digital Audio Tape systems use different sampling rates: a CD has a sample rate of 44.1 kHz, while a DAT has a rate of 48 kHz. Sample-rate conversion prevented changes in pitch and speed that would otherwise occur when transferring audio files between systems using different sampling rates.

From its origins as sample-rate conversion, resampling became a feature of digital hardware samplers "which allowed for a variety of processes to be carried out internally, with a sample processed from its original form to become 'resampled' when the process was complete" (Future Music, 2020). In the late 1980s, samplers such as the Akai S900, S950, and S1000 were endeared by electronic music producers because of the instruments' internal resampling features.[6] For example, the S950 could time stretch and "infinite loop" samples without affecting their pitch. This played a role in the evolution of jungle and drum and bass as producers created sample-based textures "that range from metallic to realistic and outright bizarre" (Fintoni, 2016). Marc Mac (4Hero), one of jungle's innovators, explained that he took the time to learn about the S950's time stretching and infinite loop capabilities by reading the instrument's manual:

> Timestretching was something like page 20 in the menu, but we needed to know everything in the sampler so that's how we stumbled on it. The other thing about the 950, that a lot of people walked away from, was how it looped. We used to call it the infinite loop. If you looped a pad sound or a drum sound or anything it would have this alternating forward and backward loop. We had a lot of people coming back to us and saying how are you making those sounds? It was simply because we didn't move quickly onto the new technology, we wanted to master the old one first.
>
> (Fintoni, 2016)

In the 1990s and early 2000s, DAW software was limited by computer CPU speed. Producers' workaround for this limitation was to resample multiple tracks of MIDI onto a single audio track in order to conserve CPU. This practice is analogous to the early days of tape recording, when Les Paul would record onto three tracks, then "bounce down" these tracks onto a fourth, freeing up the three tracks once more for additional guitar overdubs. Today, resampling is a foundational electronic music production technique. It is common to sequence a MIDI part, quickly resample it, and then alter the resampled track in various ways with processing. Alternately, the master output channel of a DAW can be resampled, as described in Ableton's Reference Manual:

> Live's Master output can be routed into an individual audio track and recorded, or resampled. Resampling can be a fun and useful tool, as it lets you create samples from what is currently happening in a Live Set that can then be immediately integrated.
>
> (Ableton, 2018: 218)

As Hertz says, resampled sounds that have been altered can be resampled again and further processed. In this way, producers' use of resampling is analogous to its origins in the sample rate conversion process: in electronic music production, *resampling converts one sound into another* through a process of layered sound designing.

Resampling is a foundational technique for four reasons. First, it's a producer's purest form of recycling creative work: folding one's sounds back upon themselves, in a feedback loop. Whether you have spent a few hours or a few weeks working on a track, the option of resampling is a reassuring reminder that you always have a trove of compelling material to play with. Second, resampling amplifies tiny sounds into larger ones. To illustrate, consider resampling a few seconds of a reverb tail. The reverb is attached to a sound that triggered it—a chorus of voices, say. Since traces of these voices are contained within the resampled reverb tail, when you resample the reverb, you also re-capture traces of the voices in that reverb-ed space. By processing the resampled reverb tail—drastically boosting its volume, or mangling it through sample-rate reduction effects—the producer discovers details or artifacts within it that were never apparent in the voices in the reverb-ed space. For Jon Hopkins, resampling is a way to boost such small overlooked artifacts in a sound:

> Boosting the little bits that you think are just noise. Boosting the mistakes. Using Ableton to really dig into the details of the sound, so I can pick out artifacts that aren't even supposed to be there. Boost them. And distort them again.
>
> (MusicRadar, 2019)

A third fact about resampling is that the producer never knows the extent of what is latent in a sound until it is resampled. Sometimes we hear in resampled sounds rhythmic things in what we thought were melodic things: formerly hidden inherent patterns and pulsations come to the foreground, shocking us into new perceptions of the music's materials. Finally, resampling is wandering, a production equivalent of straying from a predictable musical path. Producer Tim Hecker describes the excitement of discovering, early in his career, the sonic possibilities of resampling:

> taking a clinical sound out of an early software synthesizer on my computer and running it through a Turbo RAT distortion pedal, and then re-sampling it back on my computer and building something around that. It just felt like, wow, you can literally just overdrive a crappy computer pristine full-spectrum sound and degrade it. It's almost a new form of four track or lo-fi, but for me opened up all these spaces where I realized that I could work in and play in.
>
> (Burns, 2016)

Since we cannot predict what resampling will reveal, or how it might amplify tiny sounds into bigger ones and help us recycle what we already have into something new, the technique leads the music and the producer into unknown and unpredictable terrains. Resampling, in other words, is an ideal form of production Quest.

Applying effects in layers

Lessons for electronic music production disruption can be extracted from other milieus, such as visual art. Consider the drawings of Vija Celmins, who renders photographs of the ocean and the night sky. Celmins works in layers, drawing and erasing lines, building up a drawing over a long period of time and imbuing it with a sense of intention. For her, expression is the artist's thinking *compressed into the artwork through layered craft*. "All the time you're thinking and making decisions," she says. "The making, the devotion to making, is what gives it an emotional quality" (Tomkins, 2019). Similarly, the painter Gerhard Richter applies bright colors to canvases in layers, only to scrape off portions of them before adding more layers (Nowness, 2011). Artists' use of additive and subtractive processes repeated over time create beautiful and unpredictable subtleties in their works.

As a painter adds and removes layers over time, so too do producers work iteratively in layers to achieve powerful sonic effects. One approach that builds upon resampling is to process sounds with effects, one effect per resampling. For example, one layer adds reverb, and the next adds the noise of distortion. Processing effects have subtle expressive power when applied in layers: reverbs and distortions, for example, can be applied like tints and color shades. Instead of drowning a sound in a large amount of a single reverb, tint it with a small amount on one reverb setting, then tint that reverb tint with a small amount on another reverb setting in another layer of processing. Instead of distorting a sound in one pass, color bits of it in multiple passes, so that the distortions begin distorting the distortions, building ever more timbral tints into the sound.

In this way, *no single layer of effects processing is overly prominent*. The layered effects function like a patina or sonic oxidation that is the byproduct of shaping the sounds over time. As the layers accumulate, so do traces of the applied processing. This idea of accumulated layers is illustrated by the producer Daniel Lopatin (Oneohtrix Point Never). He describes his disruptive *echo jams* practice, which involves "grabbing a phrase from a track, slowing it down, and putting tons of echo on it." For Lopatin, the patina and accumulations of echoes reveal the essence of a sample and helps the producer "make this wonderful free music using organized structures" (Lopatin, 2017).

Many electronic music producers take their time experimenting with effects processing, trusting that working in layers will create new sounds in the music, new layers of evolution. The producer Holly Herndon strives to have each moment of her sounds evolve:

> Each moment of the song could have a different [effects] process. It doesn't have to be the same. It could be like moving in and out or have an interesting [effects] chain that's constantly evolving and shifting. You could automate some of it with LFOs or with automation on a timeline.
>
> (Betts, 2019)

The producer Chris Clark (Clark) processes sounds in layers, using effects plug-ins to create a textured and imperfect sound:

> I really like texture and the grain that processing puts on things. I like when stuff sounds like it's behind something, and it's not too raw. I like that sort of distance, and I don't know why. . . . I like feeling that something is a bit oblique and a bit ugly, in a way. A bit distorted

and not quite correct. For me, that's interesting, and you can only really do that with lots of processing—plug-ins. . . . Just endless processing in Ableton. Just freezing, printing, freezing that, printing that, re-pitching, transcribing that into sample libraries, bouncing that.

(Fallon, 2014)

Sleights of ear: morphing resonances

I alter the music's surface by morphing the resonances of a track's sounds. While EQing a bass part, I listen for the point at which the bass tones are most boomy-sounding so I can filter them out. Having found the boomy point, I automate the EQ so that the bass's low frequencies gradually vanish. It creates a sleight of ear: the bass loses its resonance and reappears as a treble instrument. Such *now you hear me, now you don't* morphing effects are some of the most enchanting disruptive processes in electronic music production. Morphing, producer Guy Sigsworth explains, is "placing sounds on the boundary of the real and the unreal" (Fischer, n.d.).

Sound morphing offers three lessons. First, nothing in an electronic music track need remain static. As with the example of a filter-swept acid house bass line discussed earlier, even when a part or a sound repeats itself, constant small changes to it make the repetition more engaging. Morphing shifts our mindset from thinking about production as ordering a fixed set of sounds to perpetually rearranging an ever-changing constellation of sounds. It suggests that what makes a timbre interesting is not merely its identity at one point in time, but also what that sound *does*—or has done to it—*over time*. Second, morphing invites the producer to think imaginatively about routing sounds and the nonlinear creative possibilities for disconnecting one's outputs from one's inputs. Finally, morphing shifts the producer's attention onto a track's subliminally felt presences, making manifest the many qualities that we intuit in a music *before* overtly perceiving them. Sound morphing is a subtle kind of disruption in that it makes audible almost subliminal effects that are impossible to achieve by other means.

Weird sonic artifacts: distortion

Distortion (from the Latin *distorquere*, "twist different ways") disrupts the music production process by causing sounds to behave in unusual ways. Distortion overdrives a signal so that it sounds pleasingly gritty, textured, damaged, less pristine—or what producer Chris Clark called *not quite correct*. A classic example of an overdriven distorted timbre is the amplified electric guitar sound in rock music. But any sound can be distorted, either aggressively or by subtle means, so that its timbre is either completely destroyed or evinces just hints of damage on its edges. Distortion brings to mind the Japanese *wab-sabi* aesthetic, foregrounding the imperfections of an audibly weathered sound's "raw texture and rough tactile sensation" (Koren, 2008: 68). Distortion is beautiful because it's imperfect.

One can experiment with distortions, both deliberately and accidentally, heavy-handedly and lightly, by processing sounds using various plug-ins.[7] If a distortion makes a sound more interesting to listen to, I use it. Distortion often amplifies aspects of a sound that were formerly inaudible without the effect. When distorted, a clean piano timbre becomes a buzzing-hum, as the distortion accentuates resonances in the piano's overtones. Producers often distort sounds just to hear what sonic invisibles will be foregrounded, or as Hopkins put it, *to pick out artifacts that aren't even supposed to be there*. Producer Darren Cunningham (Actress) speaks of his love for lo-fi sounds which "clashes directly with the obsession with fidelity and precision that's so common in electronic music" (Electronic Beats, 2019). For his 2008 recording *Hazyville*, Cunningham used distortions created via various sampling rates and extreme EQing to conjure what he calls a *grey sound:*

> It was all based on how those sounds were recorded, and that's how I stumbled on this strange, grey sound, because all the things that I'd sampled were different bitrates, sampled from different sources, and so to put them together you had to work in a way which balanced them against each other—EQing to the point where it lost its original form.
>
> (Lea, 2014)

When distortions on multiple tracks are added together, their cumulative timbral effect is one of depth. The music's texture becomes thicker and more intricate as the multiple distortions overlap and interpenetrate with one another to create a composite, artifact-ed sound never played but now heard and felt. Soon the music is full of inherent timbres not generated by any single track in isolation, but in combination. Over time, distortions added to parts in a track can accumulate to the point where the music takes an exponential jump in interestingness, becoming a more nuanced and complex system. We might think of layers of distortion as being a track within the track, as layers whose weathered timbres we can respond to and shape further. As music production educator Dan Worrall notes, distortion is created by adding irregular *nonlinearities* to a sound's waveform. He offers paradoxical-sounding advice to the producer wondering how to best use distortion to enrich the texture of individual sounds or a whole mix: "You need to add nonlinearities deliberately" (Worrall, 2019).

My bell part had a clean, clearly discernible sound that no longer fit with the tracks whose sound profile had evolved into murkier shapes. I needed to refine the bell sound, which meant disrupting it through distortion. I added layers of noise and weird sonic artifacts until the bell was no longer a bell. Now it sounds like an electric guitar's ringing notes feedbacking around, but not quite; it sounds almost acoustic, but not quite. Its timbre feels burdened, as if a signal is fighting the noise around it, as if trying to sing its line, but struggling to do so. As I altered the bell sound through distortion, my perception of it changed: I noticed textured and gritty shapes, ambiguous moments, random echoes and bounces, slight wobbles and out-of-tunenesses. What caught my attention were moments whose sounds surprised me because they were the opposite of a clean aesthetic. Now the bell has a sense of tension that I did not expect to find, but now that I have found it, I embrace it. In sum, weird sonic artifacts offers a production lesson, which is that *every moment in the music can be made compelling*. We aim to make music that catches our listening within a timbral counterpoint, where the details of a sound can be objects of fascination.

Conclusion

This chapter has explored the concept of disruption in electronic music production, suggesting ways for cultivating it and sharing examples of techniques producers use to advance their work. As useful as some of these techniques may be, the producer would do well to remember that disruptions are most useful when they *destabilize your understanding of what you are doing*. Ideally, disruptions change your perception by making the music momentarily unfamiliar: turning a downbeat into a possible upbeat, or transforming a bell timbre into something stranger. Changing our perceptions can happen simply or complexly, by chance or by deliberate experimentation. We conclude with producer Amon Tobin again, who neatly summarizes the point of disruptive processes: "It's all about the little psychological games you play with yourself to ensure you don't get stuck in templates or trapped in your own way of working" (Future Music, 2019).

Disruption Quest

Interrupt your track-in-progress by answering two questions:

- What could I do to disrupt the music, to destabilize what is sounding too conventional and to keep my listening on its toes?
- What would it sound like if I looped this section, muted some parts, and folded the new loop back into my arrangement?

Notes

1 As Radiohead singer and electronic music producer Thom Yorke explains, learning how a piece of equipment or workflow works means that it is no longer creatively useful: "Once you've learned to use a drum machine, or learned to write in a particular way, the temptation is to go back there, because you know it works. But . . . if you've discovered it works, it no longer works" (Frost, 2019).
2 The concept of inherent rhythms comes from Gerhard Kubik's 1962 article, "The Phenomenon of Inherent Rhythms in East and Central African Instrumental Music." Kubik describes inherent rhythms that emerge as a by-product of the patterns of *akadinda* xylophone music: "I must emphasize that these rhythms as heard do really exist and are not a product of phantasy" (Kubik, 1962: 33).
3 In his phenomenology of music performance, Arnold Berleant describes the depth of this field of action: "It is as if one were entering an immensely extended space, a space that is both fluid and temporal" (Berleant, 1999: 75).
4 In *The Creative Mind*, Margaret Boden proposes three forms of creativity, the first of which is "making unfamiliar combinations of familiar ideas" (Boden, 2004: 3).
5 (Hertz, 2016: 14).
6 For a demonstration of resampling a loop on the S950 several times in succession to create a lo-fi sound, see Repeatle (2010).
7 Popular plug-ins at the time of this writing include Izotope's *Trash 2*, SoundToys' *Decapitator*, Unfiltered Audio's *Dent 2*, and Output's *Thermal*.

7 Levels of detail
Editing

Introduction

On a 2017 Reddit discussion thread called "Tipper production methods/ethics," a producer-fan named u/coolin296 wondered about the production techniques used by David Tipper (Tipper), a reclusive and somewhat legendary electronic music producer lauded for his advanced production skills and pristine sound. u/coolin296 wanted to know more about the workflows Tipper uses "that yield such vivid, abstract, and disciplined results." A fellow Redditor and Tipper fan with the promising-sounding DJ name astral_yogi weighed in. The way to achieve what Tipper achieves, he said, was simply dedication to the craft of micro-editing:

> I read in an interview that he spends a lot of time doing what he calls "micro edits" . . . he'll take a melody and rework it every phrase with small edits to the sounds. I think that gives his music a level of detail beyond the looped phrases common in many electronic productions. I don't know if there's a secret sauce in this, just good, honest dedication to the craft. This level of editing requires patience and determination (sitting there and repeating a phrase over and over again without repeating effect patterns).
>
> (Reddit, 2017)

While Tipper himself had never discussed the topic of micro-editing specifically, in two interviews he had cryptically explained his process. His production method is simply to "sit in the studio pressing buttons until it feels like my soul is trying to climb out of its disintegrating husk. Then I know it's time to take a break. Then repeat cycle . . . Ad Infinitum" (Clash Music, 2008). In another interview, Tipper positioned his producing not in terms of specific techniques, but in terms of a self-imposed discipline required to see projects and tracks through to their completion, no matter how long such a process might take. "In order to push musical boundaries, for me at least, a huge amount of discipline is required . . . I just work on a track until I feel like it has reached its full potential. Whether it takes months or years is irrelevant" (Dodgers, 2010).

Tipper takes his time producing tracks and perhaps so should we. His patient approach to creating music over a period of months or even years is pleasingly out of sync with the notion that producers should finish tracks swiftly and efficiently, as if speed and quantity are indexes of quality. We would do well to cultivate such an equanimity for our work, to give it time to grow into the best version of itself. Besides his patient approach, little is known about how Tipper develops a track to "its full potential" by sitting in the studio and "pressing buttons." Yet producer-fan astral_yogi is probably right: Tipper's workflow, like the workflows of many skilled producers, without question involves meticulous editing.

In electronic music production, editing is the alteration of sounds after the fact of their initial improvisation, recording, sequencing, or sound design. Editing is fine-tuning sounds, but also more: choosing and moving, cutting and pasting, dragging and dropping, stretching and compressing, cleaning up or dirtying down, resampling and processing and resampling again, tightening and loosening, honing and polishing sounds to make them different from what they already are. Editing is the primary axis of producing electronic music dispersed along every stage of the process of making a track: when we produce, we are in fact editing constantly, in ways both minute and major. And as the example of Tipper reminds us, editing requires a lot of time. This chapter considers editing from several perspectives: the zoomed-in view in a DAW, as a meta-tool, as performance, as phrasing, and as a balancing of repetition with variation. It concludes by elaborating on producer Mr. Bill's idea of editing as a game of amounts.

Levels of detail

> A lot of people think it's a shortcut or a "secret" to music production and the truth is it's just spending hours and hours working on something
>
> —Ben Norris (O'Flynn)[1]

Tipper is not the only electronic music producer whose attention to editing has earned him admirers. Discussion threads about electronic music production on Reddit, DAW software user forums, and in the comments section of YouTube tutorials reveal how producer-fans listen and what impresses them the most about the better-known artists among them—from Tipper to Skrillex to Autechre. Similar to rock music fans' admiration for instrumental virtuosity such as guitar soloing prowess, electronic music producer-fans are inspired by the producer who goes to great lengths in pursuit of complexly detailed, subtle, original, and enchanting sounds. This attention paid to music production virtuosity and the dedication required to achieve it is captured in recurring online discussion thread comments—such as, *that track has an insane level of detail!*—that highlight production minutiae. While producer-fans theorize and debate such details,[2] sometimes the specifics of how a track was produced remain unclear, even to seasoned producers. In fact, part of what generates the mystique of a compelling track is the opacity of its production and an uncertainty as to what equipment, processes, and creative workflows were used to make it. Does Tipper use software or hardware? Presets or original sound design? MIDI controllers or sampled acoustic instruments? Live or programmed performances? As fans of electronic music, we wonder how Tipper made his tracks and our questions are based on our sense that the music's magic arises from its producer's attention to editing fine levels of detail into its production.

Zooming in

DAW software facilitates our working with levels of musical detail, allowing us to manipulate its minutiae. There are many tools for achieving this precision, but none more fundamental than a DAW's standard zoom function that magnifies an audio's waveform so that its micro level of detail become macro. At a zoomed in resolution, a waveform resembles a jagged, unspooled thread, and a split second of sound stretches several inches across the computer screen (Figure 7.1). Zooming in on the waveform, the producer can tidy up a volume fade, locate and mute an unwanted clicking noise, or nudge the timing of a drum hit forward or back. The power of zoom is that it allows us to see more, and thereby direct our listening onto finer resolutions as well. When the producer is done editing and returns to an un-zoomed view of the track, her

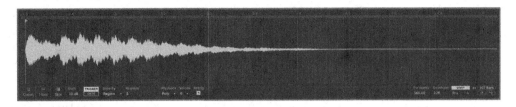

Figure 7.1 Zoomed-in audio waveform

micro-edits are invisible, yet felt: loops loop without clicks, rhythms subtly swing. After having been adjusted at a zoomed resolution, our music sounds incrementally tighter, more deliberate in its moves. *Zoom, edit, listen, and repeat.* This is progress.

If a producer spends 20% of her time *getting ideas down,* those ideas do not begin revealing their full potential until she spends the remaining 80% of her time attending to their levels of detail. As we zoom in and focus on them, minutiae we were not aware of until a few moments ago begin to appear.[3] In *Bento's Sketchbook,* critic and artist John Berger describes this interplay between noticing and a presence radiated by the work-in-progress:

> At a certain moment . . . the accumulation becomes an image—that's to say stops being a heap of signs and becomes a presence. . . . This is where your looking changes. You start questioning the presence as much as the model.
>
> (Berger, 2011: 8)

So too with electronic music production, where a heap of sonic signs becomes a felt presence and our listening shifts accordingly.

The DAW's zoom tool is a metaphor for how levels of production details work. As you produce, you zoom your attention in on an element or quality in the music, which in turn reveals itself in direct proportion to your concentration. The element or quality is anything audible: a drum sound, a reverb tail, the need for a tempo change, or a tone that is too strident. As you edit by tinkering with such details, you question the heap of signs that is becoming a presence, doing what you can to make the sounds in your track, and the track itself, come alive.

Discussion threads about electronic music production are full of knowledge, questions, and speculation about levels of detail in production practice. As we discussed in the Interlude, producer-fans are attuned to, and often skilled at noticing, details in the productions of others. On the Ableton user forum, a producer-fan named ThirdPerson wonders about how Jon Hopkins created the drum sounds for his 2013 track, "We Disappear":

> After spending a lot of time with ["We Disappear"] I've reached the conclusion that I have no idea what's going on with the drums. I know he used a lot of non-drum samples for this record (field recordings, recordings of himself tapping, etc.) but that's about the extent of what I know. My question is *does anyone have an idea what kind of processing is going on here?* And if so, how a similar effect can be achieved through Live, or if not Live, then a plug-in? Not trying to bite his style, but I'm sincerely clueless about how these sounds were made and just curious and interested in the process.
>
> (Ableton, 2015)

Drawing on their own listening to Hopkins' music, other Ableton users try to answer ThirdPerson's question. A producer-fan named Airyck explains the processing effects he hears on "We Disappear":

> I think what's happening here is a fair amount of bit crushing, sample rate reduction, and heavy compression/sidechain compression used creatively to cause the whole thing to move together in a rhythmic fashion.
>
> (Ableton, 2015)

Another producer-fan named morgo suggests that ThirdPerson read a 2013 interview with Hopkins in XLR8R magazine, where "he speaks about his production process, including a really interesting working technique with Sound Forge." In the XLR8R interview, Hopkins had referenced "the sheer detail of editing and rhythmic stuff" he could generate using Sound Forge, an audio editing software program (also used by the producer Burial, as discussed in Chapter 5) in which he could experiment with "an infinite number of changes on a sound." In addition to editing, Hopkins used Audio Ease's convolution reverb, Altiverb, to amplify a sound's otherwise latent timbral details in the hopes of happening upon something compelling:

> Some of my favorite plug-ins, like Altiverb, can be used for so many things because you've got all these models of spring reverbs and such in there. You put that on a sound, bounce it, slow it down, and you get these incredible harmonics. I've found a lot of weird sounds through processes like that. These really beautiful melodic accidents that I try to capture.
>
> (Fallon, 2013)

Such discussions about how electronic music producers create or discover their sounds show electronic music production as an ever-changing field of action shaped by producers' relationship to, and understanding of, their gear coupled with idiosyncratic processes and stylistic goals—such as Hopkins' search for "beautiful melodic accidents." Online discussion about unusual or compelling sounds connects producer-fans with producers who make widely circulated and admired music. Producer-fan talk and interviews reveal both amateurs and professional producers alike as students of electronic music production continually on Quests to zoom in on new ways of making music.

Levels of details within a level of detail

Using the Zoom tool, there are, additionally, many levels of details within a level of detail. Over many sessions, I have been shaping percussion sounds—boosting, cutting, effecting, and carving the boom *kicks*, snare *snaps*, and hi hat *chics* so that they sound more interesting on their own and more cohesive as drum kits. There is no end-point to this tinkering except the point at which a sound *feels just so*.[4] During one session, I use Live's Erosion effect to add white noise to the percussion (Figure 7.2). As I play with the noise's shape and texture by adjusting the shape of the effect's band-pass filter, I hear new details arising within this level of detail.

The added white noise becomes a new timbral layer in the percussion sounds that helps differentiate them from other sounds in the track. I adjust the settings on this new layer, moving around the band-pass filter to make a darker or brighter sound, then automate it so that it changes over time. Making such small changes to the percussion by adding a white noise layer has me

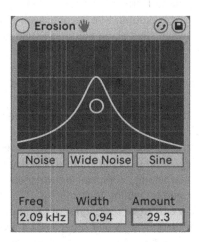

Figure 7.2 Ableton Live's Erosion effect

thinking about other elements in the track. Let us zoom through descending resolutions of attention at play in the production of the music so far, from macro to micro, from a wide level of detail to an ever narrower one:

- the track
- the percussive parts of the track
- the kick and snare parts of the percussion
- the timbre of the kick and snare parts
- the high frequencies of the timbres
- the white noise added
- the band-pass filter controlling the white noise
- the settings on the filter
- the movement of the filter's settings over time

Editing these levels of detail within a track's parts makes subtleties within them more audible and subliminally felt. Through editing, the producer continues to notice the music's levels of detail in the search for new ways to accentuate and highlight them.

Editing as meta-tool, as performance

The editing practices of electronic music producers offer models for how we might work. The producer and visual artist Carsten Nicolai (Alva Noto) is an illustrative starting point because he habitually uses a radically minimal sound palette in his music.[5] Nicolai explains the centrality of sound editing to his production process and how editing itself, rather than using a varied palette of sounds, is his main tool:

> I tend to use instruments that I am used to using, having worked with them for a long time. I think every musician has their favorite tools and you have to understand that more complex

instruments require more time to learn, and you have to put the time in to learn them. For me, one of my main tools now is editing. I've become a very fast sound editor. When trying to solve a problem, I will always think about editing a sound rather than trying to recreate it.

(Nicolai, 2018)

For Nicolai, editing is the most fundamental meta-production tool for altering and manipulating all aspects of one's sounds.

When a track is underway and initial parts have been committed to, our editing moves respond to the changing needs of the music. There are broad categories of action through which producers edit their work, including: arranging/re-arranging, shortening/lengthening, interrupting/disrupting, warping/morphing/blending, fading in/out, amplifying/distorting/filtering, rhythmifying, reverbing/delaying, softening/sharpening, compressing/expanding, inverting/reversing/mirroring, accelerating/slowing down, and tightening/loosening.

By incorporating these broadly construed techniques of manipulation into our production process, we can shape each part of the music so that it becomes a presence for the duration of the track that dynamically interacts with the sounds around it, approximating the emergent flows of a real world complex system. As we try out various sound editing techniques we wonder, *Is this helping the music be more expressive?* Editing is often laborious, but time spent patiently listening and re-listening to our experiments inevitably leads to discovering new presences to notice in the track, new layers of complexity to which to respond.

One way to think about editing is as a kind of performance. In essence, many of the electronic music producer's editing techniques are after-the-fact, virtual versions of what a musician does when playing an acoustic instrument. Just as skilled performers use their instruments to both produce sounds and control them to make the music come alive, so too does the producer use editing to achieve something similar. Even though editing happens iteratively and over a period of time outside the duration of a track, it is similarly concerned with manipulating the music so that it appears to perform itself.

Some producers devise ways to perform editing in real time while they are improvising and recording their music. For example, producer Daniel Lanois describes a workflow he used while working with Brian Eno in the 1980s, whereby signal and effects processing such as reverb and EQ "were recorded during the performance, not added during mixing time" (McNeill, 2015). Lanois explains the ebb and flow virtues of *performing with effects* rather than setting them to fixed levels. This, he says, makes effects *interesting on the spot*:

> I appreciated that Eno included me in the creative process: that he encouraged me to perform on the graphic equalizer as he was performing on the synthesizer, which gave the performance an ebb and flow; a sonic ebb and flow, not an aestheticized synthesizer sound. I've realized the importance of putting a sonic character into the recording, not to just record something flat with the view of making it more interesting later. To make it interesting on the spot, be it by printing your sound effects or working a graphic equalizer as somebody is performing.
>
> (McNeill, 2015)

The producer Ryan Lee West (Rival Consoles) does a similar kind of live editing while recording initial compositional ideas he plays on hardware synthesizers into his DAW, as we saw in Chapter 4. For *FACT Magazine*'s *Against The Clock* series on YouTube, West builds a track in ten minutes, using a Sequential Circuits Prophet synthesizer routed through several guitar effect pedals. As West plays chord clusters on the synthesizer with one hand, he simultaneously adjusts the instrument's controls and settings on the effect pedals using his other hand, thereby *editing the sound signal before it arrives into the DAW*. In a Reddit AMA, West explains his method:

> I also record in all the synth sounds live, with some performance of the filter, volume, lfos changing etc. on the synth itself. Then out of this they often go through some guitar pedals such as boss od3, boss DD3, Moog midimurf, lofi junky Z.vex, roland space echo 555, tape delay.
>
> (West, 2018)

Editing as phrasing: micro-information and subliminal feeling

Editing is also a way to refine a music's phrasing by allowing the producer to re-contextualize concepts learned from playing acoustic musical instruments. My own experience as a percussionist illustrates this carry-over from the acoustic to the digital. When I studied western classical percussion, my lessons frequently touched on ways of phrasing the music to make it more dynamic through sticking patterns, shaping passages from soft to loud or loud to soft, and shifting tempos. Such phrasing concepts are relevant to editing in a DAW, where a MIDI representation of a performance is a world of control information, displaying the contours of a melody line's rise and fall, timing, and the velocity level (volume) for each note.

As I listen to the music while looking at it, I notice phrasing problems with my recorded performance: *notes become softer as the melody ascends, and louder as it descends*. The MIDI velocity levels confirm what I see. Dragging thin vertical lines up and down, I reverse the dynamics of the phrase, so that it gets louder as it ascends, and quieter as it descends. It sounds more sensible this way, meaning: how I would play it had I played it right. Editing dynamics is the simplest of editing moves, yet massively impactful in clarifying a music's phrasing, allowing the producer to create a sense that a track's sound and the gestures that produced it are aligned.

The producer can think about performing phrasing in the context of every parameter of a track, from its velocities to its melodies, from its beats and effects to its arrangement and mix. Whether improvised live or recreated after the fact through automation, phrasing conveys micro information like a subliminal messenger, slinging our attention around the music's various depth layers and helping the sounds convey themselves just so.

Editing the relationship between repetition and variation

In electronic music production, especially dance-oriented EDM styles, repetition and variation are inextricably connected. Whether it's a steady four-on-the-floor kick drum thumping, arpeggios in hocketing synchrony, flickering loops of percussive noise, or overlapping layers of ambient pulsations, repetition generates the groove aura of a track, drawing us into the enchantment of the music, towards a trance sensation. Some musics single-mindedly pursue this trance goal with hard-hitting beats in the tradition of the disco to techno dance music continuum, while other musics use less obvious forms of oscillation and pulse. But no matter what the musical style or what form of repetition we work with, variation is a way to complement repetition's relentlessness. Like a kaleidoscope that allows us to view changing color patterns, variations foreground repetition's beauties while moving the music forward.[6]

Since electronic music producers tinker with variations of as many parameters of the music they can automate, mouse click on, and get their imaginations around,[7] one principle for creating such variations is avoiding repeating musical elements whose sound does not change in some way. As producer Ben Lukas Boysen observes: "a question that will always be part of a musician's work is if what he creates is contributing to what's already there or if it's an unneeded repetition" (Fischer, 2014). Thus, a general production rule of thumb: *If you repeat a sound, vary it*. A sound that sounds itself in predictable ways, over and over again, soon becomes uninteresting, and often annoying. Inventive producers find ways inside repetition to conjure a sense of direction and organic variations. TJ Hertz describes the task of producing electronic music as "how to make it sound as if it's going somewhere" (Martin, 2018). Similarly, producer Alejandro Ghersi (Arca) explains the importance of incorporating high levels of detail and variation into tracks to emulate an "organic" sound:

> I very painstakingly spend time embroidering such a degree of detail and variations, because that's my way of attempting to emulate something that's more organic. I feel like it never loops the exact same way twice or three times. I'm an enemy of something repeating in the exact same way or too long, because you lose access to a particular kind of feeling of unpredictability or discomfort.
>
> (Morse, 2015)

For Ghersi, painstakingly embroidering the music through editing details and variations is a way to *hypercharge* the listener's attention so that they feel the music differently. Ghersi's words recall John Berger's notion of a work transforming from being a heap of signs to being a presence that we notice anew. "I think if you hypercharge some of the attention span, then the attention span is almost overwhelmed" Ghersi says, "and a trap door opens up underneath it and enters something deeper" (Morse, 2015).

I shape my drum sounds, one by one, aiming for them to have a sense of musical line, a sense of phrasing, a sense of surprise, and repetition with a purpose. Here and there I re-pitch sounds along the drum parts in my arrangement, tuning the kick drum to make a bass line. My sound-shaping changes how I hear the sounds, shifting my listening stance from annoyance to curiosity, from *Oh, it's that sound again* to *Oh, what's that sound?* The more I edit the drum hits, the more what I hear engages me. It is also becoming clear that the track could benefit from many more such adjustments. Like zooming in on ever finer levels of detail, I extend my editing from drum sounds outwards to the many other elements of the track, adjusting timbres, tuning, timing, and transitions between sections. Through editing, the producer aims to craft perceptual trap doors that may open up something deeper in the music.

Conclusion: editing as micro-adjustments and the game of amounts

Summing up editing we borrow an idea from the producer Mr. Bill, who describes production as *a game of amounts*. Mr. Bill is referring to the multitudes of micro-adjustments made in the course of editing the parts and sounds that make up a track. When you are producing, you are in fact editing continuously: editing is improvising, tinkering with a musical system, designing and re-designing sounds, shaping beats and adding noise to them, folding ideas back in upon themselves, arranging, mixing, and re-mixing. Electronic music production is a game of editing over time with the goal of getting the music sounding just so. Trust your sense that what you are making is either exciting or boring, a recycling of old gestures or a sound that is genuinely new to you (for now).

Sometimes in production, no matter where we are along the process of building a track, we have the solutions to your questions in front of us. As you listen to your music-in-progress, you need only notice a single thing that you did not notice yesterday. It can be as simple as a level that is too high, a moment where the energy drags, or a sound that needs finessing. Once you identify something in need of improvement, try out ways of improving it. This is editing as a game of amounts.

Editing Quest

Your editing Quest is to revisit a production project you have already begun and adventurously tinker with its sounds. To edit, zoom in on the details of the MIDI and audio and change them. Tighten the timing of a beat to perfection or loosen it to total slack, incorporate more volume fades, add phrasing, re-pitch, get precise with your effects processing. Take a cue from the producer Arca: edit a degree of detail and variation into the parts until each one sounds *hypercharged*.

Notes

1 (Payne, 2019).
2 For example, see the 2018 Reddit discussion thread, "Who are the most technically talented producers in EDM/electronic music?" (Reddit, 2018).
3 "Surprising detail," notes John Salvatier in his essay on craft, perception, and self-improvement, "is a near universal property of getting up close and personal with reality" (Salvatier, 2017).
4 Skilled musical tinkerers, notes Steve Waksman in his study of heavy metal guitarists, have "mastered the subtle process of making technological adjustments to achieve a desired result" (Waksman, 2004: 697).
5 For example, in his collaborations with Ryuichi Sakamato, Nicolai has worked with a sound palette consisting of mostly sub-bass and high-pitched sine tones.
6 Jean-Michel Jarre speaks to this idea of variations on repetition:

> When I started to do electronic music I was obsessed . . . about not having anything being repeated in exactly the same way. For me it was exactly the opposite attitude to that of Kraftwerk, Tangerine Dream and all those electronic bands who were doing something more robotic. I considered electronic music in a much more sensual, organic way, where nothing should be repeated (Flint, 2008).

7 Automation is the movement of a software device's controls or parameters over time. The Ableton manual describes automation thusly: "Often, when working with Live's mixer and devices, you will want the controls' movements to become part of the music. The movement of a control across the song timeline or Session clip is called *automation*; a control whose value changes in the course of this timeline is *automated*. Practically all mixer and device controls in Live can be automated, including the song tempo" (Ableton, 2018: 303).

8 Conjuring

Arranging and mixing

Introduction

The music is coming along. You have 10, 20, or maybe 60 tracks of audio and MIDI, bits and pieces of beats, bass, chords, melodies, samples, and atmospheres layered together. You have improvised and recorded parts, disrupted and edited sounds, routing them from here to there, processing them to sculpt their details. A final stage of the process is arranging and mixing the music to balance its tracks so that the bits and pieces enter and exit at the right time and sound right. An arrangement is the overall structure of a track, and for most producers who use a DAW to produce music, the craft of arranging is putting all of the music's layers into a horizontal form. A mix is the overall sonic balance of a track, and the craft of mixing is balancing the dynamics and frequencies of individual parts into a vibrant composite stereo whole. In electronic music production, mixing and arranging are inextricably intertwined: changing a part in your track's arrangement can change how its mix sounds, and vice versa. This chapter considers some general concepts that inform these tasks.

Arranging and arrangements

On Reddit in 2019, a producer named dj-Nate asked the electronic music production community for advice on how to build an engaging arrangement. Feeling trapped in formulaic structures, dj-Nate wondered about how to best organize a track to keep listeners engaged and shared a breakdown of his typical go-to arrangement for dubstep tracks:

> I hope this isn't a dumb question. For this post I am mainly describing my dubstep arrangements. I feel confident in my sound design and mixing, but I tend to get trapped creating arrangements like so:
> 30–40 second intro
> 10–20 second buildup
> 1min-1min 30 second drop
> Copy entire song so far, paste after drop
> Modify copy/pasted intro to become a break
> 10–20 second buildup #2
> Modify drop to be different
> 10–20 second outro
> How can I make my arrangements more unique and memorable? What are some tips to keep the listener engaged?
>
> (Reddit, 2019)

Fellow producer-fans responded with suggestions that encouraged dj-Nate to think anew about the elements of his arrangements. A reader named Bleeyl pointed out that while dj-Nate's conventional arrangement structure is useful, another way of approaching the problem is to focus on *micro-arranging* elements within the arrangement by addition or subtraction:

> Your arrangement doesn't need to be unique/memorable . . . that structure sounds fine 2 me, and is a structure that most songs follow. Also, there's a lot to be said about the more micro-arranging that goes on, especially when we're talking vertical arrangement: What elements? Do we start minimal/melodic which adds and adds or with a full groove already going which breaks down? Questions like that which you already tackle when you're writing all the time, even if you're not straying from your conventional macro-arrangement that make your song unique regardless.
>
> (Reddit, 2019)

Another producer-fan, 7rlpLvk3ly, suggested using one's favorite tracks as arrangement models:

> Try lifting arrangements of tracks you find compelling. Load them in your DAW, set tempo, and use track markers. Listen to 4 and 8 bars at a time on a loop and take notes. Do this a few times and you'll have a better idea how to create them yourself.
>
> (Reddit, 2019)

Although focused on dubstep music production, such producer-fan discussion highlights how a producer might think about a track's arrangement. In the case of dj-Nate, perhaps part of the quandary was his searching for an overly simple and definitive answer to what is an open-ended question. In electronic music production, there are many ways to arrange a track without relying on EDM's DJ-oriented conventions of intros, builds, drops, and outros. A potential problem with such pre-fab arrangement structures, notes producer Barry Lynn (Boxcutter), is that they risk being "geared towards the goosebump moment that doesn't actually give you goosebumps, because it's too contrived" (Gibb, 2011). Like improvising, sound design, rhythm programming, disruption, or editing, arranging is vast world of possibilities compressed into decisions about what should happen in a track and when.

Arrangement zoomed out

On Twitter in 2020, the producer Ryan Lee West (Rival Consoles) tweeted a zoomed-out screenshot of the Ableton Live project file for one of the tracks on his 2018 album, *Unfolding*. West, whose music has been described as "high-brow, avant-garde electronic music" (Buerger, 2018), showed the arrangement of "Hidden," 7 minutes of pulsation overlaid with piano chords and atmospherics. The screenshot West tweeted was an impressive vertical display, showing some sixty tracks of audio unfolded like a long scroll. West explained his arrangement:

> There are a lot of improvised layers recorded in on different channels which come together to create waves of movement. pretty much everything is audio recorded in and then arranged. Many small, little sounds all very simple, but together create the image.
>
> (West, 2020a)

In a follow-up tweet, West explained that while the screenshot of his track's arrangement appears dense, the music has space: "note this is 7+ minutes across so actually it is far more sparse than this image suggests." Most of his production time, West noted, "was spent deciding the overall structure" (West, 2020a).

West's sharing his Ableton screenshot and his explanation of its contents was an unusual (and generous) gesture that few producers would reciprocate on social media, perhaps because most consider their DAW project files guarded artistic and intellectual property. And since a project file displays a trove of information about a producer's decision-making—from editing to automation to effects routing to overall structure—one could perhaps sleuth out something from another producer's work simply by reverse-engineering from a screenshot of their track.[1] In any case, West's producer-fans were grateful for his having shared the photo and impressed by the arrangement it displayed. Even at a zoomed-out resolution, West's project suggests the considerable level of detail involved in producing a supposedly "sparse" track by arranging its many parts into layers whose audio "image" has a tapestry-like structure.

Such discussions about arrangement on Reddit and Twitter highlight how there are no magic formulae for building an effective arrangement. While dj-Nate sought advice for how to make an arrangement organized in timed sections more unique and memorable, West showed a different way of thinking about a track as an organized weave of parts.

Arrangement as Lego blocks

In a DAW, a producer can create an arrangement by moving MIDI and audio objects around the timeline of the Arrangement page, as if the objects are Lego blocks of sound, modular visual patterns "that can be measured against a time grid" (Zagorski-Thomas, 2014: 135). Sometimes, this drag and drop approach can lead one to predictable musical structures, and for this reason some producers find looking at their music displayed linearly to be problematic. Mark Fell advises us, "if you are using a screen, turn it off and reach out with different forms of cognition." Fell contends that one downside of the DAW is the Arrangement page's linear display of musical time, which narrowly defines the producer's arranging task. "In effect, your role becomes one of organizing bits of information in a grid like space" he says, "where you can see the beginning, middle and end, and all the layers running alongside one another" (Fell, 2016).

Like Fell, Kieran Hebden (Four Tet) reflects on the limitations of the DAW's visuality. In an interview in 2003, when laptop-based electronic music production was just beginning to be a viable practice, Hebden explained how seeing parts vertically line up in the timeline shapes how a producer hears the music:

> People who make music on computers don't realize how powerful the visual element is. Whether you like it or not, your minds starts to think in terms of patterns, because it's a natural human way to do things, and you start seeing the way drums are lining up on the screen, and it becomes completely instinctive to line them up in a certain way. It's important just to close your eyes and use your ears, and trust what's coming out of the speakers more than anything.
>
> (Inglis, 2003)

Yet other producers use the DAW's default timeline to their advantage. Producer Sara Abdel-Hamid (Ikonika) builds arrangements from a starting point of an "overloaded 8 bar loop" which she then deconstructs and spreads out over time:

> When I make a tune, I will concentrate on 8 bars at a time. But that first 8 bar loop has to bang for me. . . . What I actually like to do is have an overloaded 8 bar loop and then spread it out across 3 to 4 minutes or whatever, and get that arrangement done.
>
> (Pazienti-Caiden, 2020)

At the end of a YouTube tutorial on arrangement, producer Nolan Petruska (Frequent) similarly explains his technique of finessing details within 1-bar sections at a time. Petruska's aim is to eventually duplicate the 1-bar sections along the timeline to create longer 4-bar sections that will incorporate subtle sound design evolutions:

> Essentially what we want to do is take each of these parts and . . . have each of those be a four-bar section, and then find a way to evolve it slowly and subtly. . . . Now we're into this Jon Hopkins-, Floating Points-inspired type thing, so it needs to be this longer, drawn out sort of arrangement.
>
> (Petruska, 2020)

Producer James Parker (Logos) also arranges his music by positioning its elements along his DAW's Arrangement page:

> Often when a track's 70% complete it won't really sound finished and everything comes together in the edit at the end. So I don't work in a real time way—I [use] a drum and bass way of producing, which is working in the timeline, positioning things in that four- or five-minute timeline, then going back and forth and doing the structure that way.
>
> (Wilson, 2019)

For his 2013 recording, *Cold Mission*, Parker arranged the music by experimenting with different positionings of *sonic events* on the Arrangement page, carefully choosing which of its few sounds occur, where they occur, and why:

> I was really focusing mainly on music where there's—and this is to an extent what happened on *Cold Mission*—very few sonic events, so you have to choose quite carefully what they are; and you're thinking quite carefully about where they're positioned; and you're experimenting with what effect they have on you as a listener.
>
> (Lubner, 2019)

Arrangement as narrative and expression of complexity

The options as to how to arrange a track are evocatively explained by producer and organist Kara-Lis Coverdale, whose music has been described as having a "narrative dynamism" (Ballard, 2017). For her, an arrangement "is a way of unfolding narrative in music to produce the musical depiction of an event, an adventure or a special feeling" (Jackson, 2016). Recalling our discussion in Chapter 1 about musical systems being complex systems, Coverdale makes an analogy to the complex systems of the natural world, which are "beyond the possibility of complete knowing while at once being unmistakably perfect in alignment" (ibid.). Ideally, an arrangement is a similar expression of complexity whose perceptual effect enchants even the producer herself. An arrangement, she says, is a balance point that "teeters between what is mindfully constructed and also makes even me wonder how I made it" (ibid.). Evoking composer Arvo Pärt's idea, discussed in Chapter 2, that our creative task is to find the "appropriate system for the gesture," Coverdale thinks about arrangement as *inherent* in the sounds of one's materials:

> I think the arrangement is at once determined and unpredictable at the strike of the first note. The first note already contains so much arrangement-relevant information: context, purpose, history/origin of the instrument or sound and character. Is it a muted sound or a

certain damaged, gentle threat? Is it percussive and lively or sporadic and full of energy? Is its decay quick or sustained?

(Jackson, 2016)

Finally, Coverdale also hears a track's arrangement as an opportunity to transcend "whatever compositional framework you are using because that in itself suggests (if not demands) an arrangement formula." In lieu of a formula, the producer's arrangement task is *figuring out how to best organize what is already latent in the music*. This holistic idea of arrangement as a structure-to-be latent in one's sounds reframes dj-Nate's intro-build-drop-out arrangement conundrum. Perhaps Coverdale's perspective would help him hear ideas already in his intro section that could be organically expanded to structure the entire track? In sum, most producers would agree that the most impactful arrangements, however structured, sound *inevitable* and allow the component parts of a track to "express themselves clearly and support the others in the whole" (Jackson, 2016).

Arrangement as transitions, effects fills, effective automation, and incidentals

I listen while looking at the shape of the arrangement on the screen, scrutinizing the transitions more closely. Now that I have a rough arrangement taking shape, how one section connects to the next comes into play and transitions are the means of this connecting. My track's sections are easy to see on the screen, but ideally they would not sound that way. In some styles of electronic music, dramatic drum fills, FX build ups, risers, temporary part muting, and bass drops are useful transition devices that let the listener know what is coming next. In his YouTube tutorial on arranging a techno-style track, producer Brent Sadowick (Sadowick) frames the task thusly: "How are we going to make transitions? How are we going to make it more palpable? How is the audience going to anticipate something that is changing? (Sadowick, 2014). But in other musical styles, producers devise ways for their tracks to transition from one section to another without being obvious about it. Sometimes the producer's puzzle is figuring out how to have the music be without perceptible seams so as to sound like a continuous flow.

One way of connecting two sections of music together is via a *common tone*, such as a pitch that is part of two different chords, or a leftover, non-pitched sound that connects by sustaining long enough, such as a decaying cymbal crash or the resonance of a reverb's tail. Another connecting technique is sharing a melody, chord progression, bass line, or rhythm among different parts, from one section to another. Such a shared musical element carried over a transition point from section to section creates continuity and flow. An example might be a hi hat whose rhythm reappears in a bass line. Such transitions can be used to connect even disparate ideas into a subtly interconnected arrangement.

Many tried and true production techniques for connecting come from the DJ's heritage, such as the fade in/fade out using volume and EQ controls. A DJ gradually brings up the volume fader of the incoming recording to blend it with what is currently playing, and then fades out the currently playing one—making one sound disappear while a new one emerges. A related technique to the volume fade is EQ filtering, whereby low, mid, or high frequencies are momentarily removed from the mix. Filtering frequencies is a way to play with the timbral shape of the music and create drama: cutting mids and highs, for example, muffles the music, which creates anticipation for the imminent return of what Sadowick called "something that is changing"—in this case, the missing frequencies. A producer who uses these mixing techniques to reign in an arrangement's world of possibilities approaches a track as a DJ might—as a sequence of sounds that enter and exit the mix. For TJ Hertz, an arrangement "largely amounts to effective automation and bringing parts in and out." He advises producers to "be patient with the [parts] that are there and give them space to breathe" (Hertz, 2016).

Finally, impactful arrangements often rely on single sonic events known as *fills* or *incidentals* that connect parts, bridge sections of a track, and fill out the mix. Jon Hopkins calls this type of transition *effects fills*:

> What I'll do right near the end of the production is more of the morphing type sounds that you hear bridging sections together; similar to drum fills except they're effects fills. That can really bring the music to life, but I'll only do it early on if a track is dragging in a section.
>
> (Future Music, 2018a)

Hertz similarly speaks of the importance of production *incidentals*. Like the producer-fan 7rlpLvk3ly who offered dubstep arrangement advice to dj-Nate on Reddit, Hertz recommends listening to the structure of one's favorite tracks as an exercise, paying attention to how small events subliminally impact the music's unfolding:

> Try loading your favorite tracks of a similar genre into a DAW and annotating every single event, no matter how minor: "incidental here", "filter cut-off rises on lead", "reverb swell", "hi hat decay opens up" etc. You might be surprised at how much "little stuff" is actually contributing to the overall flow of the track and how few "major" parts there actually are.
>
> (Hertz, 2016)

Case study: Autechre's "bladelores" (2013)

In some electronic musics, a track's arrangement is deeply intertwined with its sound design in that both involve subtle and continuous change to sounds over time. The complexities of such arrangements are well illustrated in the information-dense music of Autechre. Let us consider a bird's eye analysis of the 12-minute track "bladelores" from Autechre's 2013 recording, *Exai*, using a play-by-play account of the music's process and structure to get a sense of the track's arrangement.

"bladelores" begins with a funky muted kick drum and a simple white noise backbeat on 2 and 4 that is drenched in reverb and joined by a repeating acid-style bass line. The groove almost sounds like slow hip hop. At 1:00, a pulsating harmonic texture joins the mix, blending in with the long reverb tail triggered by the white noise backbeat. At 2:16, the pulsating texture becomes louder, accentuating the offbeats. The groove continues, but what seemed to be a reverb tail has morphed into a kind of chordal wash that is steadily growing. Around 3:15, we notice the chordal wash is in fact two chords that are alternating and repeating, and by 4:00 the backbeat is fraying a bit, while the bass line's timbre becomes more squelchy through a filter opening and closing, in the style of a Roland TB-303 sequencer. At 4:55, the chords and backbeat cut out, leaving only the squelchy bass line, which now feels brittle.

Soon—from 5:11 to 5:37—the chords surge to the foreground again for a moment, hitting a kind of resolution, only to cut out at 5:38 where the backbeat, reverb tail, and bass line return, now slightly altered. The chordal wash joins in again at 6:35 and for the next two and a half minutes grows in intensity as the percussion and bass line flutter about. The reverb from the outset of the track has been transposed onto the chords, making their resonance grow to gargantuan proportions. At 9:00, the backbeat abruptly stops, leaving the bass line to slowly dissolve into the resonant chords that continue thickening until they become a resounding wall of harmonic sound at 11:00, and then gradually fade out as the track ends.

This is the rough arrangement structure of "bladelores." The music's details, which are not easily renderable in prose, manifest themselves as changes that happen to the music in a continuous flow.[2] For example, in any one-minute segment of the track, any individual sound—the beat, bass line, reverb tail, or the chords—reveals micro evolutions inflicting themselves continuously on it, second by second. Thus, the white noise back beat is never *only* a marking of beats 2 and 4, nor is the bass line merely marking a chord progression. In Autechre's music, the parts keep changing rhythmically and timbrally, and this change is the basis of the arrangement's process which creates "a now whose content changes ceaselessly" (Burrows, 1997: 529). In sum, "bladelores" flows in a way that recalls Coverdale's ideal arrangement form, which is *at once determined and unpredictable at the strike of the first note.* Unlike a cookie-cutter song form that risks sounding cliché, the track does not proceed from one section to another, but instead shape-shifts over time. The arrangement is a presence in the shape of a process, as if the music is an evolving organism.

Arrangement Quest

Build a 3-minute arrangement from the smallest possible musical idea—a beat, a chord, a sample, or wash of noise. Use repetition, cutting and pasting, and DJ-style fade in/outs to create an arrangement with a sense of dynamism and inevitability.

Mixing

> The mixing process can sometimes take months for a single track and requires hundreds if not thousands of micro-decisions about every sound in a track.
>
> —Steve Hauschildt[3]

The above quote from producer Steve Hauschildt captures the time-consuming complexities of mixing, which touches every component of electronic music production. A mix combines multiple individual tracks of audio into a single stereo composite, balances all of the music's timbral textures (e.g. hard and soft, boomy and brittle, hazy and clear, dry and wet), creates a sound object with energy and enchantment, emphasizes the important parts at each moment, exaggerates or compresses dynamics, generates ambiance, unifies contrasting sounds and stratifies similar ones, focuses musical center points and edges, and positions foreground and background. But practically speaking, a mix answers the question, *How can everything be heard?*

Mixing as conjuring

To begin, consider a remote example: Hieronymus Bosch's evocative 16th-century painting, "The Conjurer" (Figure 8.1). In the painting, a man performs a magic trick for an assembled audience. In his left hand he holds a basket, in his right, a small ball. On the table in front of him are the tools of his trade: a few balls, a stick, and two upturned cups. One astonished onlooker leans in to look more closely at the ball held by the conjurer, as if thinking, *How did he do that?*

From magic trick to audio track: Bosch's "The Conjurer" could serve as a metaphor for the mix and for the sleights of ear involved in mixing music. The producer as Conjurer: imagine the Conjurer's two upturned cups as studio monitors emitting stereo sound to the wonder of the music's listeners. An enchanting mix—the sum of what Hauschildt calls *thousands of micro-decisions about every sound in a track*—is ultimately a conjuring sleight of ear that creates the impression that we are hearing more than we are, that somehow the music extends beyond what we are able to notice, as if it is manipulating our perception, putting us under a sonic spell. Jay

Figure 8.1 Hieronymus Bosch, *The Conjurer* (1505)

Hodgson elegantly encapsulates what is at stake for producers mixing their music: "recordings don't reproduce sounds. They model hearing." Mixes, he says, "are the 'conjuring trick' engineers use to generate the auditory delusion of musical activity within a single acoustic phenomenon" (Hodgson, 2017: 222). In *Mixing Audio*, Roey Izhaki connects this conjuring to sonic depth, noting that as listeners we "are used to sonic depth in nature and we want our mixes to recreate such sense. The main objective is to create something natural or otherwise artificial but appealing" (Izhaki, 2017: 72). By what means does a mix sound natural or otherwise artificial but appealing? And what can electronic music producers do to move their track mixes towards this enchanted state? This final section considers some perspectives on these two questions.

A mix as a landscape with four dimensions

Imagine the mix of your track as a journey through a landscape in sounds that unfolds over time, as you, the listener, ride through it on the vehicle that is your listening. You notice many spatial details and relationships as the landscape wizzes by. In the foreground, the small trees that line the road are a blur of activity. In the distance, rolling hills and forests pass more slowly. And far on the horizon motionless mountains meet sky. The whole scene is awash in sunlight, but it's late in the day and there are a few scattered clouds about, so the light is changing subtly, bright over here but casting shadows over there. When you look around from the perspective of your moving self through the landscape that is the music, you notice multiple things happening on the different timescales of foreground and background. Taken together, the passing road, small trees, rolling hills, the sky, and play of light create a sense of ever-changing spatial depth in the visual mix you experience as you move through it.

With this image of the mix as a landscape, your most impactful production move is to exercise as much control as possible on as many sound qualities that you can perceive in your mix as it unfolds over time. Your aim is to create the impression of the music extending beyond what is sounding, creating a sense of enchantment. An enchanting mix conjures the sense of 3D depth analogous to what you would experience as you moved through a landscape. Producers create this depth by crafting a sense of perspective, whereby some sounds are close up front and some far at the back, some in the center and some off to the sides, some sounding quite clear and present while others more indistinct and faded. Much of this depth can be created through careful adjustments to the volume, EQ, compression, and panning of parts.[4]

This idea of a mix as a journey of sounds in four dimensions, passing through the ongoing moment of listening, is commonly referred to as the mix's *soundstage*, which has four components: stereo width, frequency height, timbral depth, and positional movement. Let us briefly explore each component.

The first dimension is the *left to right stereo field*. Producers position or *pan* sounds anywhere along this field, from the far left to the far right, or anywhere in between. But sounds can also move around. While a kick drum or a vocal is often positioned in the middle of the field, other percussion or FX sounds might move out to the sides and then back towards the center of the mix, creating a sense of motion. The producer can automate such panning movements in a mix.

The second dimension of the mix is its *top-to-bottom frequency spectrum*. High frequency sounds like hi hats we hear as located high up or "on top" of the mix, low frequency sounds like kick drum and bass tones we hear as located down below, and mid-range frequency sounds like marimba as sitting in the middle. While these sound types are not literally vertically positioned (only metaphorically so), each one occupies a different stratum of the mix's overall frequency spectrum. Mixes with clarity designate and carve out a frequency stratum for each sound, so that different sounds do not clash with one another. Martijn Van Sonderen, of the production trio Noisia, explains the importance of picking and shaping a track's musical elements so that they fit within the mix's frequency spectrum and stereo field:

> The most important thing is to pick your elements. There's no point just randomly throwing all kinds of sounds together and then trying to blow everything up. You have to deal with the limitations you have within the spectrum and stereo field. For individual elements it's good to check what frequencies are not necessary for this particular sound to work in your mix, so you can EQ them out.
>
> (Murphy, 2016)

The third dimension of a mix is its *front to back depth*. The main determinant of how close a sound appears to us is its volume, but timbre and ambiance also impact this perception. Sounds without ambiance such as reverb or delay (i.e. "dry" sounds) and high frequency sounds generally appear closer and more present, while sounds with ambiance (i.e. "wet" sounds) and more lower frequency content appear further away. In general, notes William Moylan, the amount of "subtle low-level information" contained in a timbre determines how far away or up close a sound appears to us in a mix. Moylan reminds us that sounds "can be crafted to sound astonishingly close to the listener, or in infinitely far" (Moylan, 2017: 37). The mixes of producer Jon

Hopkins create a sense of this proximity and vivid, front-to-back depth. Hopkins explains how he experiments with settings on reverb plug ins, playing with this subtle information to

> create very strange effects where things just sound really set-back, and you don't know why. . . . I use a lot of room and phase effects, not to make things sound wide or narrow, but for head-fuck sounds where you can't tell where they're coming from. There are a lot of great plug-ins for that these days.
>
> (Smith, 2018b)

The final dimension of a mix is its sense of *motion over time*. Any sound can come and go in volume, move about the stereo field, ascend and descend the frequency spectrum (to some degree), and move from a far back sense of depth to an up close sense of proximity. Such movements imbue the mix with a sense of motion that is independent from whatever rhythmic motion is already happening in a track through repeating rhythms, interactions among parts and within the arrangement, and pulsations from effects such as delays. There are, then, a plethora of moving elements to consider as one constructs a mix.

The electronic music producer attends to the four dimensions of the mix with the goal of making the music more enchanting, meaning that its sounds create for the listener the sensation of hearing and feeling things that are not obviously audible, and experiencing a composite sound that appears more than the sum of its parts. A mix is thus like a *trompe l'oreille*, or ear-fooling, a realistic conjuring.[5]

Listening contexts: lessons from mixing on headphones

It is worth untangling the perceptual advantages and limitations of mixing on headphones, since many electronic music producers use them while working on laptops in mobile studios. The main advantage of headphones is that they remove variables in the acoustics of far-from-perfect home studios that distort how we hear a mix. As engineer Andrew Scheps puts it, with headphones "you're basically wearing your studio on your head" (Scheps, 2019). Another advantage is that, with their speakers positioned on our ears, headphones enable us to hear levels of detail within the mix's component sounds—from the grain of a reverb to the tone color of subtle EQ adjustments. But headphones have limitations with regards to how they distort our perception of low frequencies and the stereo field's spatial qualities. Most headphones do not realistically reproduce low frequencies because these are slow moving and require a large physical space in which to build up and be heard accurately. Case in point: it is difficult to *feel* bass through headphones. Another distortion of headphones is due to their binaural design. The coupling of left and right channels to our left and right ears prevents the natural psychoacoustic phenomenon known as audio interference or *crosstalk* that happens when we hear some of the sound from the right monitor with our left ear, and some from the left monitor with our right. It is this lack of crosstalk on headphones—the minute time and frequency differences in sounds as they reach our ears caused by the *head shadowing* effect—that prevents us from experiencing the realistic sense of immersive stereo space conjured by monitors. On headphones, mono sounds seem to originate inside our heads rather than externally, the center of the stereo image appears empty, and parts panned to the extreme right or left can sound disconnected from the rest of the mix.

In sum, although mixing on headphones is perhaps not ideal for the reasons just mentioned, it is nonetheless useful because it teaches the producer to listen closely whilst being mindful of the complex variables that shape how we perceive sound. Some electronic musicians embrace mixing on headphones. Producer Rob Clouth mixed his immersive 2020 recording *Zero Point* on headphones, using their stereo separation to spur experimentation with binaurally oriented panning techniques. Of the mixing process, Clouth notes, "I could hear everything absolutely perfectly; there was no background noise or anything" (Freeman, 2020). Thus, while headphones introduce their own distortions, as we learn to identify these distortions we gain another perspective on mixing. Try, for example, listening to your favorite producers' tracks on your headphones and compare them to your own. "The whole point," Scheps reminds us, "is to have a listening environment where you can be creative and your mixes translate to whatever system somebody might listen to your mix in" (Scheps, 2019).[6]

Balancing objective listening with subjective feeling

Ryan Lee West's tweeted photo of his Ableton project file showed the music's arrangement, but of course this image does not explain how the producer went about mixing the track, which would have been fascinating to learn about.[7] How did he mix "the many small, little sounds?" We don't know. But we can use our not-knowing productively. As we work on our own mixes, we can learn to recognize the feedback loop between the ever-changing sound of the music and our ever-changing perception of that sound. As we listen, questions about our tracks arise: Is the mix cluttered, or have I not found the right arrangement for its component sounds? Is that part too loud or problematic for some other reason? When we mix, we hear not only the music but also how we listen.

I'm listening through mixes of *Plentitudes*, beginning to hear more of what is missing and more of what to listen for—a sound that is too loud or soft, a boxy or strident or thin or boomy timbre in need of EQ, an obvious (or not obvious enough) reverb tail, a mix that is dull and needs more high frequency "air." I trust my sense that noticing qualities in the mix has been earned from spending time identifying problems and trying to improve upon them. But as I listen it is clear that things *not* in the mix are equally problematic: *my listening is unstable*. A fact of music production—indeed of all musical listening—is that when we listen we are also *listening to our interaction with the object of our listening*. This idea is described by Thomas Clifton, who says in his classic phenomenology of music, *Music as Heard*, that as we listen we are "observing the self observing the music" (Clifton, 1983: 22). Listening to my mixes is listening to both the sounds and to myself engaging with the sounds. It is, then, difficult to know what objectivity I have about what I hear. *Sometimes the problems are in the sounds, and sometimes with me.*

As producers mixing our own music, it is worth reflecting on our positioning with regards to what we are hearing. Recalling our discussion of mindset in Chapter 3, what is your vantage point?

- Do you listen from a point of doubt or sympathy, skepticism, confidence, or anxiety?
- Does your positioning allow you to hear the music as it is, or as you wish it would be?
- Do you have a sense of the music's big picture mood (as it might be apprehended by a non-music producing friend), or are you caught up in its minute technical details?

As we mix we are at the same time playing with different mixes of our noticing, as if moving its faders to alter our perceptions and maybe compensate for our aural blind spots.

Mixing a track with many parts leads us to notice how and what we hear, because mixing requires us to discern different musical lines and timbres simultaneously and figure out ways to balance them. As we find a rough mix among parts, we begin noticing new details within this balance, which in turn inspires new mix adjustments. Soon our noticing is, to quote Arca again, *hypercharged*, and we scramble to adjust to what we hear. But even though the music is moving closer to how we want it to sound, it can be beneficial to maintain a degree of doubt. Do we notice what is most prominent or what we are most attuned to?

The electronic music producer's tracks are measured and calibrated in precise ways, a compressing of many parts into a unified whole. At the same time, the music is an accumulation and documenting of whim, accident, and improvisation. As we mix we alternate between two very different modes of being, toggling between being critical and being supportive of the track, being analytical and associative, being among the details of the parts and at a distance surveying their overall impact. Balancing these different mindsets requires keeping our thinking robust despite our flickering perceptions and passing moods. In sum, any producer benefits from the constant interplay between striving for some kind of listening objectivity and an awareness of inevitable listening subjectivity. One way to achieve a balance between objective and subjective listening is to *take notes while listening to a mix* to identify problem spots with a degree of lucidity. TJ Hertz, a producer of pristine mixes, explains the technique: "As you listen through, write down things that don't quite sound right, and the corresponding time in the track." With this list, return to your DAW file and address the problems one at a time. Once this fixing is completed, bounce down another mix and listen again. "The really important thing," he says, "is that every idea that you act upon must first be written down. That way, you don't lose track of what you set out to achieve" (Hertz, 2016).

Less is more: muting, reduction and efficiency

The mixes are coming along—far along, considering where the music was a few months ago—but I want them to surprise me more, to disorient me. There could be more enchantment with where and when each sound enters the mix, and also with how the arrangements develop. I had preconceptions about how the tracks would build and sound, but now? Now there is too much. Somewhere along the way, I forgot about the power of emptiness.

As discussed in Chapter 6, a way to do more with less is to mute parts in a track, clarifying a musical texture by silencing bits of it here and there. "It's what you don't hear, more than what you hear, that makes something clear," says Serum software designer and producer Steve Duda. "It's the absence of junk. . . . If you can shorten sounds, the more you're going to create space for the next thing to be interesting. I'm almost always shortening sounds" (Duda, 2015). The producer can apply muting to every part in a track, like a cleaning. As we mute notes or sections, striking musical textures can arise: moments of beauty reveal themselves, the arrangement gains space, and correspondingly, the mix becomes clearer. As producer Ben Lukas Boysen notes, each element in an arrangement and its mix "should have a meaning, function and reason to be there" (Fischer, 2014).

A mix compresses time into a track

I return to each track every few days, tweaking the mixes bit by bit, spreading out a workflow. A track contains many layers worked on over time, *compressing the time of its making into its sound.* As we work iteratively on our mixes, time spent allows us to refine its sounds. If a track's enchantment is proportional to how it has been manipulated, calibrated, and finessed over time, then an enchanting mix surely contains traces of every moment spent making it, in the form of a gradual accumulation of layers. Layers of what, exactly? Layers of sound, but also layers of thinking and decision-making, layers of fretting and hoping, layers of doubt and wrong turns, and layers of amorphous thoughts that appear whilst listening repeatedly to a track in progress. Hertz again:

> My working style is one of revision after revision. . . . It's like chipping away at a block of stone and as you progress you get stranger and stranger shapes and as soon as you try to carve one thing you'll suddenly see different faces
>
> (Courant, 2014)

For Hertz and probably many producers, production is ultimately "about making sense of the music that comes from this process" (Courant, 2014).

Conclusion: six general production concepts

This chapter has described some general approaches to arranging and mixing in electronic music production. These tasks are interconnected because decisions in both domains affect how sounds play out over time. While there are excellent resources for learning their fundamentals (e.g. Felton et al., 2018), there is no better teacher than diving deep into producing one's own music by taking the time to experiment with its form and sonic textures. As a guide to organizing a track's sounds so that they conjure maximum enchantment and perceptual magic, we conclude with six production concepts that inform effective arrangements and mixes.

Contrast

Bright sounds let you hear the dull sounds, low sounds let you hear the high ones, full textures let you hear the sparse ones, slow lets you hear fast, and so on. Contrast allows all manner of musical differences to be articulated and heard.

Compensations

When one part comes up, something else comes down, or else everything will be coming up and competing with everything else for your attention.[8] Even if it is only for a passing moment achieved in a subtle way, have one musical thing foregrounded at a time.

Musical lines

Each part can have its own arc over the piece, a journey it goes on. Even if the listener doesn't consciously notice this line, it is nevertheless felt.

Surprise

The way to make the music un-boring is through change, disruption, and surprise. Try not to bore yourself.

Efficiency

What is the least I can do in the mix that makes the maximum impact?

Accumulations

Individual mix changes made over time will accumulate as layers and thereby accrue a composite power.

Mixing Quest

As with every other aspect of electronic music production, self-imposed constraints help focus our attention on creating a mix that sounds articulated, sensible, and alive. Your mixing Quest is to constrain your options for a production project by making adjustments to track volume levels and EQ (equalization).

First, using the volume faders on your DAW's mixing page, balance the track levels so that each track is loud enough to be heard, yet not so loud that it "sticks out" (If you find a part sticking out, it is likely too loud. Or the part itself could have musical problems). If needed, automate the levels of tracks so that parts "listen" to one another by making space for their respective entrances, or otherwise draw your attention to the music's changing point of focus. In the days before mixing automation, recording engineers would *ride the faders* of their mixing consoles to manually adjust such volume shifts in a final mix. Now you can draw in automation for these subtle cues of attention.

Once you have made adjustments to levels and set automations, use an EQ plug-in—sparingly!—to more precisely sculpt the mix's component frequencies. EQ parts to give each one its own frequency stratum, to remove non-essential frequencies (e.g. the low frequencies on a high-pitched sound) and make space for other parts, to highlight parts that are not sufficiently audible, and to finesse timbres. If you use a quality set of headphones with a flat frequency response, trust what you hear: Does a part sound bright? Dark? Muffled? Strident? Experiment with different EQ settings to shape the sound you want. Locate frequency ranges in need of boosting or cutting. For example, give a sound more "air" by increasing the volume of its high frequencies using a shelving setting to boost a broad band, or reduce a bass sound's boominess to carve space for a kick drum.

Through careful adjustments to volume and EQ, make your mix a vivid, articulated presence.

Notes

1 An earlier example of such a photo is a screenshot from one of Autechre's Digital Performer project files that appeared in *Sound On Sound* magazine in 2004. The photo caption: "A forest of Digital Performer automation and MIDI controller data gives some clue as to the detail that goes into Autechre's programming and sound editing" (Tingen, 2004). On his website, the producer Mr. Bill sells his Ableton Live project files for download. He writes: "Play, modify, reverse engineer, enjoy!" (Day, n.d.).
2 Paul Morley speaks of our "attempt to match the otherness of music with an otherness in the writing about it" (Morley, 2005: 328).
3 (Hauschildt, 2019).
4 Tutorials on mixing topics can be found on YouTube. An authoritative book on mixing is Roey Izhaki's *Mixing Audio* (2017). William Moylan's *Recording Analysis* includes in-depth discussions of sound-staging and the illusion of space on recordings (Moylan, 2020: 287–356). For an introduction to mixing fundamentals, see White (2006).
5 Autechre member Sean Booth explains his interest in making mixes less obvious – meaning, without all of their sounds being hyper-present, or positioned "up front" of the mix's soundstage. He uses the phrase *deep mixing* to describe this sonic conjuring:

> This tendency to make everything sound upfront – if you were to look at a YouTube mixing tutorial, the tendency is to make every sound audible. I've gone back on that, to the point where I'm really interested in deep mixing now, where you've got things you aren't necessarily aware of at first listen. Having a sound stage that's got almost like a 3D feel, that's the only way I can describe it . . . (Sherburne, 2018).

6 At the time of this writing, there are numerous headphone correction plug-ins such as Waves' *Nx – Virtual Mix Room*, Goodhertz's *CanOpener*, Sonarworks' *Reference*, and Slate Audio's *VSX* that aim to (1) make mixing on headphones feel more realistic by emulating the crossover sound of listening on monitors and (2) conjure the acoustics of various listening environments. Slate Audio's software, for example, models spaces including "LA Nightclub," "Boombox," "Electric Car," and "Howie Weinberg's Mastering Room."
7 There is, however, a video of West generating track ideas for FACTmagazine's "Against The Clock" series (FACT Magazine, 2016).
8 This impossible mix situation was once humorously described by Deep Purple guitarist Ritchie Blackmore, who allegedly asked a mix engineer: *Can I have everything louder than everything else?*

Conclusion

The push forward

This book has considered facets of the craft of electronic music production by connecting a variety of perspectives from historical sources, the talk of producers and music production producer-fans, and from this author's own experience. We began with a brief history of music production, focusing on the figure of the producer, the emergence of MIDI, sampling, and the DAW, and examined musical systems and workflows. Then we considered components of the production process: ways of beginning a project, using presets and designing sounds, improvising and being lost, rhythm programming, disruptions and editing, and finally, arranging and mixing. These activities are not always separate; in fact, they often overlap with, and interpenetrate, one another as producers mix while layering parts, edit as they sound design, arrange as they improvise, and so on. Moreover, there are as many ways of producing electronic music as there are practitioners of the craft. Most producers develop, over time, unique musical systems as well as idiosyncratic ways of using these systems through their workflows. This book has been concerned with asking broad questions and seeking perspectives on electronic music production as a field of knowing and modality of tacit knowledge. Admittedly, production's horizons extend far beyond the examples discussed in this book, not to mention its author's know-how.

In the Introduction, I described producing as a Quest that involves seeking out workflows for making music whose approach, methods, and craft create sense from your musical system and, ideally, make sense of the totality of your musical life. The Quest of electronic music production is figuring out ways to push your craft forward with the tools you have and the attention you devote to your work. The pushing forward concept is key, because it reminds us to hear beyond the particularities of our equipment and *learn how to think with and through it*. "The push forward," says producer Amon Tobin, "is in your own mind, not the gear you've got." Tobin explains that the way to advance one's productions is through conscious and thoughtful decision-making:

> The only things you can really take credit for, creatively, are the decisions you make. It's not so much the instruments you play or how good your voice or technique might be, but what you've decided to leave in or out—the creative editing process in your mind. It makes no difference to me whether you're finding sounds in a synthesizer or guitar; your creative contribution is really what you've decided to keep and how you structured the melody and rhythm.
>
> (Future Music, 2019)

This chapter pushes forward by bringing together some of the book's broad themes into a final mix of suggestions for how to cultivate and curate creativity in electronic music production. I consider how to tighten the production feedback loop, suggest seven production principles, examine failure, and finally, encourage the pursuit of authenticity.

Infinite perceptual games: four lessons from tightening the production feedback loop

While working with DAW software and other digital equipment may seem like a deeply mediated, and at times, alienated experience, producing electronic music is in fact a bio-feedback journey in that it offers repeated opportunities to get in tune with our learning in a state of deep engagement, to shape ourselves through the shaping of our tracks. As we play, program, record, edit, and mix, we become *attuned*, encountering opportunities for tightening the connection between what we are doing and what we are learning from that doing. Consider, then, four lessons from this feedback loop.

Make many small errors in a tinkering-based process of experimentation

The virtue of making small errors, notes the mathematician and essayist Nassim Taleb, is that they do little harm yet "are also rich in information" (Taleb, 2014: 21). In electronic music production, our errors appear in the form of trying out sounds and processes, over and over again, until we encounter something compelling. It is not unusual for producers to explore dozens of sound types, processing techniques, or layering combinations in search of a composite timbre that feels just so and produces a kind of perceptual magic. When the only harm that comes from such extensive trial and error is spent time, that production time is well spent.

Work on the thing that is most problematic or in need of altering

Trying to fix what needs fixing is the quickest route to improving our productions. Listen closely to the parts and hone in on a single sonic thing—an errant frequency, one too many kick drum hits in a beat pattern—and assess how you might improve it in a concrete way. Since each sound in your track makes a competing claim on your attention, practice focusing your listening on what most needs your help. With experience, our listening perceives more as we turn up the volume of our attention like faders on a mixer, separating the music's signals from its noise.

As you alter, fix, or improve something, take note of the specifics of what you are doing

Pay attention to your actions—including apparent errors that did not lead you where you wanted to go—in order to learn from them and incorporate their lessons into future projects. "In the end, the existing flaws of one composition are the potential benefits of the next one," notes producer Ben Lukas Boysen (Headphone Commute, 2016).

As you work on the music over time, let your changing perception of it feed back into you process

What sounded good to you yesterday can sound uninteresting today because you have learned through the experience of producing. Trust that what you notice now is an accumulated sense. You are trying to refine the music so that it works today, tomorrow, and in the future, so be generous with it by bringing all of today's know-how to bear on making it cohesive, expressive, efficient, and able to hold your attention. As you work to improve it in ways that you notice and feel, you are sharpening yourself too.

Seven principles of electronic music production

One of my own discoveries over the course of writing this book is that there are many ways to arrive at a production goal, many ways to create a track. As I finished the *Plentitudes* project, I learned that I have few methods of producing more useful than trying out many different things and pursuing those that sound the most compelling. In lieu of methods, then, consider seven principles by which to push forward one's productions.

Begin by playing something and capturing it

This is as simple as it sounds! Play . . . anything. Improvise on an instrument: play chords on the keyboard, drum a beat for a bit, record a soundscape. Trust that a moment of your performance is enough to build on. The important step is the act of Capture: recording yourself playing or triggering a sound, or recording the world outside the studio. In your playing and capturing are traces of ideas that are not yet fully formed or apparent to you. No matter how brief, a performance is the DNA for a future track, containing ideas that you can explore, play with, and develop.

Develop something simple by making it more complex

Now that you have captured some kind of performance, build on it. You can build on it in time, by extending it horizontally, or you can build on it in timbre, by expanding it vertically. To extend your performance, copy/stretch it in whole or in part, so that 1 measure becomes 10 or 20. To complexify a sound, effect it, layer other sounds with it, or resample it to make a different-sounding copy. Through such techniques, what began as something simple becomes more complex.

Notice, keep going, and keep noticing

Having played something and begun the process of making it more complex by building on it in time or in timbre, notice what you now have and keep tinkering with it. As you develop complexities or pare down into simplicities you will notice new sounds, new levels details. Maybe a sample shard you repeated makes an interesting rhythm, or a series of effects you layered creates an unusual ambiance. *Pay attention to what you are noticing now.* The more time we spend producing, the more we notice that attention is the most important part of our musical system because attention connects the possibilities of production tools with our imagination.

Refine

As you notice new sounds emerging, refine them. The aim of refining your sounds is to make their essence more articulate, or, if the situation requires it, less articulate. A clear bell sound can be made clearer, a buzzing drum sound can be buzzier. Refining your sounds means making them more of what they already are, pushing them to extremes, sharpening them. We produce by responding to small sonic things that we notice and organizing them further still. "A creative process," as Wayne Bateman noted in *Introduction To Computer Music*, "involves a progression from a state of relative disorder to an advance state of higher order" (Bateman, 1980: 246).

Reduce, arrange, and mix

Reduce what you have by removing inessentials. On *Plentitudes*, I removed many parts I had laboriously constructed (and I could have removed more) because the music simply sounded better without them. Arrange and mix the music by playing with what is left to create a sonic balance that makes sense.

Assess quality

Asking *Do I like this music?* moves your listening from focusing on fine details back to the big picture. Maybe the track is finished, or maybe it still requires work. Maybe it is the first in a series, or practice for something else. Whatever the case may be, you have moved the music along and learned more about production in the process.

It takes time

The craft of producing is a gradual process. Creating and developing parts, noticing subtleties, refining, reducing, arranging, mixing, and assessing quality takes time. So take your time.

Failure

Woven into any finished track are small failures that we do not know how to fix, perhaps because they come from the very manner in which we work. In electronic music production, there is failure of imagination, failure of execution, failure to creatively use our production tools, and failure to arrive at Perfect.

Failure of imagination includes failure of conception or not adequately thinking through and committing to a vision for the music, and failure to take risks. Failure of execution is slapdash editing, unclear relationships among parts, and lifeless arrangements and mixes. Has a track ever been compromised for you because some aspect of it that was not adequately thought through? This is a failure of both imagination and execution. Conversely, how enchanting it is to hear a track that is carefully crafted with attention to its details! As we work on our music, we try to make it fail less by making its components incrementally more sensible and satisfying to hear.

There is also our failure to creatively use our musical systems. The potentials of production tools such as the DAW and its associated software plug-ins set a high bar in terms of what we *can* do but are probably *not* doing. Simply put, we are *never making full use of our musical system*. At any given moment there may be many potentially more interesting yet unsounded sounds we did not pursue. With each discovery we wonder, *What else have I missed?* Electronic music production is endless in its creative routes that will never be pursued. But recognize what you can and cannot control. Failures of technique, form, or mixing can be improved upon, while other kinds of failure are simply workflow realities to be aware of.

A final by-product of producing music is that each project feels like a failure to arrive at Perfect. The producer is nagged by the possibility that a finished track *could have been better*. But there were too many variables, unknowns, shortcomings of the music's maker, and not enough time to get everything just right. We are left with Imperfect music—one more iteration that moves us forward on our production Quest. This is the nature of the production process. Now onto the next track.

Final thoughts: curate your own authenticity

The electronic music producer works within the constraints of limited resources and time, aspirations and ambitions, and anxieties about fitting in. But within these constraints there is always space for curating authenticity.

A starting point is *pursuing only that which sounds compelling*. Compelling musics keep us returning to them because they make us feel something and we hear new things each time we listen. How you arrive at compelling is irrelevant; what is important is how it sounds. While there are established conventions of production—e.g. *don't make a track too long, don't over-compress the sounds*—sometimes these conventions ways merely map what has already been done, what has already worked for others within an accepted production paradigm. If you need to, sidestep trends long enough to try out your own producing ways.

The next step for curating authenticity is to *build upon whatever you find compelling by devising your own workflows for developing your material*. The technologies of production are such that *there is no end for ways of combining them* to create a complex musical system that inspires you. New combinations and configurations of software and hardware tools *always* lead to new sounds. Your Quest is seeking out production practices that resonate with you.

A third step: *don't try to fit it*. If only for a moment, suspend your disbelief about the music's prospects of finding listeners. The producer John Frusciante provocatively frames this step as a question to consider: "What can I do that would be totally divorced from the effort to please people with music?" (McDermott, 2020). Focus on the processes of your own production practice, move at your own pace, pursue your own enchantments, and answer your own questions.

Repeat these three steps: follow what sounds compelling, develop it through new workflows and configurations of your musical system, and refrain from trying to fit in. As you practice producing, the music becomes more authentically yours with every iteration. In sum, electronic music production is a field of tacit knowledge in which we continually ask questions through sound, work with sound as a complex and dynamic system, and welcome the uncertainties and enchanting happy accidents along the way.

Suggested listening

Chapter 1

Four Tet, "Two Thousand and Seventeen"
Les Paul, Mary Ford, "How High The Moon"
The Beatles, "Being For The Benefit Of Mr. Kite!"
Phil Spector, The Ronettes, "Be My Baby"
The Tornados, "Telstar"
Glenn Gould, Johann Sebastian Bach, *The Well-Tempered Clavier, Book 1*, "Fugue No. 20 in A Minor, BWV 865"
King Tubby, "Clock Face Dub"
Ted Macero, Miles Davis, "In A Silent Way"
Pierre Schaeffer, "Chemin de Fer"
Roni Size & Reprazent, "Brown Paper Bag"
Boris Brejcha, "Lost Memory"

Chapter 2

Flying Lotus, "Zodiac Shit"
r beny, "Alone in the Pavilion"
Nosaj Thing, "Medic"
Kaitlyn Aurelia Smith, "Existence in the Unfurling"
Caterina Barbieri, "Fantas"
Pole, "Überfahrt"
TM404, "202/202/303/303/606"
Lanark Artefax, "Corra Linn"
Mark Fell, "Multistability 6-B"
Aphex Twin, "minipops 67 [120.2] [source field mix]"
Terry Riley, "In C"
Steve Reich, "Come Out"
Brian Eno, "Discreet Music"

Chapter 3

John Cage, "In A Landscape"
Biosphere, "Angel's Flight"
Akira Rabelais, "Then the Substanceless Blue"
Deru, "1979"

Rival Consoles, "Forwardism"
Stenny, "Detraction"
Imogen Heap, "Hide and Seek"
Arvo Pärt, "Fratres"
Harold Budd, "The Serpent (in Quicksilver)"
Harold Budd, Brian Eno, "First Light"
Keith Jarrett, *Köln Concert*
Squarepusher, "Midi Sans Frontieres (Avec Batterie)"
SOPHIE, "Pretending"
Ólafur Arnalds, "ypsilon"
Objekt, "Silica"

Chapter 4

Justin Bieber, "Running Over"
Depeche Mode, "See You"
Laurel Halo, "Raw Silk Uncut Wood"
Joris Voorn, "Mano"
Amon Tobin, "One Shy Morning"
Skee Mask, "Rev8617"
Matmos, "No Concept"
Tim Hecker, "That World"
Loscil, "Angle of List"
Boards Of Canada, "Turquoise Hexagon Sun"
Philth, "Let You Fall"
Ital Tek, "Leaving The Grid"
Mr. Bill, "Apophenia"
G Jones, "Everything All at Once"
deadmau5, "Glivch"
Steve Hauschildt, "M Path"
Ben Lukas Boysen, "Love"
Guy Sigsworth, "North"
Autechre "32a_reflected"

Chapter 5

Jean-Michel Jarre, "Oxygene, Pt. 4"
Afrika Bambaataa, The Soulsonic Force, "Planet Rock"
Public Enemy, "Don't Believe The Hype"
New Order, "Blue Monday"
J Dilla, "Laser Gunne Funke"
The Winstons, "Amen, Brother"
Roni Size & Reprazent, "Brown Paper Bag"
Timbaland, Justin Timberlake, "Cry Me A River"
Harold Melvin & the Blue Notes, "The Love I Lost"
Giorgio Moroder, "I Feel Love"
Jon Hopkins, "Open Eye Signal"
Jlin, "Enigma"

Nils Frahm, "#2"
Burial, "Archangel"
Floating Points, "Bias"

Chapter 6

Skrillex, "Scary Monsters and Nice Sprites"
Phuture, "Acid Tracks"
Ricardo Villalobos and Max Loderbauer, "Reblazhenstva"
Logos, "Cold Mission"
William Basinski, "Tear Vial"
Au5, "Neptuna"
Culprate, "Inside"
Sam Barker, "Models Of Wellbeing"
Lorenzo Senni, "Superimpositions"
4Hero, "Loveless"
Oneohtrix Point Never, "Power Of Persuasion"
Holly Herndon, "Fear, Uncertainty, Doubt"
Clark, "Legacy Pet"
Actress, "I Can't Forgive You"

Chapter 7

Tipper, "Gulch"
Alva Noto, "Uni Rec"
Arca, "Time"

Chapter 8

Boxcutter, "Frame Stack"
Ikonika, "Not Actual Gameplay"
Frequent, "Broken"
Kara-Lis Coverdale, "2c"
Autechre, "bladelores"
Noisia, "Collider"
Rob Clouth, "Casimir"

References

Ableton (2015). Jon Hopkins' Drums – Processing, Method, etc. *Ableton*. Available from: <https://forum.ableton.com/viewtopic.php?t=214939> [Accessed 18 November 2020].

Ableton (2018). *Ableton Reference Manual Version 10*. Available from: <https://cdn-resources.ableton.com/resources/0b/c1/0bc1007e-bd0b-4d6f-b52d-ee0054f3a6f8/l10manual_en.pdf> [Accessed 18 November 2020].

Adria, F. (2010). *A Day at el Bulli: An Insight into the Ideas, Methods and Creativity of Ferran Adria*. New York, Phaidon.

Akrich, M. (1992). The De-Scription of Technical Objects. In: Bijker, W. E., Hughes, T. P. & Pinch, T. eds. *The Social Construction of Technological Systems*. Cambridge, MA, MIT Press, pp. 205–224.

Amendola, B. (2010). Jimmy Bralower: Web Exclusive. *Modern Drummer*. Available from: <www.moderndrummer.com/2010/12/jimmy-bralower/> [Accessed 18 November 2020].

Ampong, D. (2018). Programming Real Feel into Drum Machines. *Waves*. Available from: <https://www.waves.com/programming-real-feel-into-drum-machines> [Accessed 18 November 2020].

Anthony, T. (2019). Studio Talk: Matmos. *Electronic Musician*. Available from: <www.emusician.com/artists/studio-talk-matmos> [Accessed 18 November 2020].

Appleton, J. & Perera, R. (1975). *The Development and Practice of Electronic Music*. New York, Prentice-Hall.

Attack magazine (2013). Andreas Tilliander. *Attack Magazine*. Available from: <www.attackmagazine.com/features/interview/andreas-tilliander-tm404/> [Accessed 18 November 2020].

autotranslucence (2018). Becoming a Magician. *Autotranslucence*. Available from: <https://autotranslucence.wordpress.com/2018/03/30/becoming-a-magician/> [Accessed 18 November 2020].

Avanti, P. (2013). Black Musics, Technology, and Modernity: Exhibit a, the Drum Kit. *Popular Music and Society*, 36 (4), pp. 476–504.

Baccigaluppi, J. & Crane, L. (2011). Brian Eno: U2, Coldplay, Talking Heads. *TapeOp*. Available from: <https://tapeop.com/interviews/85/brian-eno/> [Accessed 18 November 2020].

Bach, G. (2003). The Extra-Digital Axis Mundi: Myth, Magic and Metaphor in Laptop Music. *Contemporary Music Review*, 22 (4), pp. 3–9.

Ballard, T. (2017). Kara-Lis Coverdale: Grafts. *Pitchfork*. Available from: <https://pitchfork.com/reviews/albums/23224-grafts/> [Accessed 18 November 2020].

Barry, R. (2017). *The Music of the Future*. London, Repeater Books.

Bateman, W. A. (1980). *Introduction to Computer Music*. New York, John Wiley & Sons.

Beatnick, M. (2015). Interview: Floating Points. *FACT Magazine*. Available from: <www.factmag.com/2015/10/16/floating-points-interview/> [Accessed 18 November 2020].

Bell, A. P. (2018). *Dawn of the DAW: The Studio as Musical Instrument*. New York, Oxford University Press.

Bencina, R. (2012). The Interview: A Conversation with Roger Linn. *AudioTechnology*. Available from: <https://issuu.com/alchemedia/docs/at90> [Accessed 18 November 2020].

Bennett, S. (2019). *Modern Records, Maverick Methods: Technology and Process in Popular Music Record Production 1978–2000*. New York, Bloomsbury Academic.

Berger, J. (2011). *Bento's Sketchbook*. New York, Verso.

Berleant, A. (1999). Notes for a Phenomenology of Musical Performance. *Philosophy of Music Education Review*, 7 (2), pp. 73–79.

Berlina, A. (2015). Art, as Device. *Poetics Today*. Available from: <https://warwick.ac.uk/fac/arts/english/currentstudents/undergraduate/modules/fulllist/first/en122/lecturelist2017-18/art_as_device_2015.pdf> [Accessed 18 November 2020].

Betts, W. (2019). Holly Herndon Thinks Electronic Music Should Sound Like it's 'Alive and Breathing'. *MusicTech*. Available from: <www.musictech.net/features/interviews/holly-herndon-ai-electronic-music/> [Accessed 18 November 2020].

Bevan, T. (2018). Serum Tutorial – Organic Synths Like Flume/Louis the Child/Mura Masa. *SynthHacker*. Available from: <www.youtube.com/watch?v=EDISgpWtnzM> [Accessed 18 November 2020].

Bijker, W. E., Hughes, T. P. & Pinch, T. eds. (1987). *The Social Construction of Technological Systems*. Cambridge, MA, MIT Press.

Blanning, L. (2016). Playing by the Rrefoules: Creativity through Constraints. *Ableton*. Available from: <https://www.ableton.com/en/blog/creativity-through-constraints/> [Accessed 18 November 2020].

Boden, M. (2004). *The Creative Mind: Myths and Mechanisms*. London, Routledge.

Bonnet, F. J. (2020). *The Music to Come*. London, Shelter Press.

Brejcha, B. (2018). Putting the Base of the Song Together: Boris Brejcha on Producing Minimal Techno. *Cubase*. Available from: <www.youtube.com/watch?v=lK_vOMnwKDI> [Accessed 18 November 2020].

Brend, M. (2012). *The Sound of Tomorrow: How Electronic Music Was Smuggled into the Mainstream*. New York, Bloomsbury Academic.

Brett, T. (2015). Autechre and Electronic Music Fandom: Performing Knowledge Online through Techno-Geek Discourses. *Popular Music and Society*, 38 (1), pp. 7–24.

Brett, T. (2016). Virtual Drumming: A History of Electronic Percussion. In: Hartenberger, R. ed. *The Cambridge Companion to Percussion*. Cambridge, Cambridge University Press.

Brett, T. (2017). Polyrhythms, Negative Space, Circuits of Meaning: Making Sense Through Dawn of Midi's Dysnomia. *Popular Music*, 36 (1), pp. 75–85.

Brett, T. (2019). Popular Music Production in Laptop Studios: Creative Workflows as Problem-Solving within Ableton Live. In: Hepworth-Sawyer, R., Hodgson, J. & Marrington, M. eds. *Producing Music*. New York, Routledge, pp. 179–193.

Brett, T. (2020). Prince's Rhythm Programming: 1980s Music Production and the Esthetics of the LM-1 Drum Machine. *Popular Music and Society*, 43 (3), pp. 244–261.

Brovig-Hanssen, R. & Danielsen, A. eds. (2016). *Digital Signatures*: The Impact of Digitization on Popular Music Sound. Cambridge, MA, MIT Press.

Buerger, M. (2018). Rival Consoles: 'Unfolding.' *Pitchfork*. Available from: <https://pitchfork.com/reviews/tracks/rival-consoles-unfolding/> [Accessed 18 November 2020].

Burgess, R. J. (2014). *The History of Music Production*. New York, Oxford University Press.

Burgess, R. J. (2020). Random Access Changed Everything. In: Zagorski-Thomas, S., Isakoff, K., Stévance, S. & Lacasse, S. eds. *The Art of Record Production: Creative Practice in the Studio, Volume 2*. London, Routledge.

Burns, T. L. (2016). Tim Hecker. *Red Bull Music Academy*. Available from: <www.redbullmusicacademy.com/lectures/tim-hecker-lecture> [Accessed 18 November 2020].

Burrows, D. (1997). A Dynamical Systems Perspective on Music. *The Journal of Musicology*, 15 (4), pp. 529–545.

Buskin, R. (2007). Classic Tracks: The Ronnettes' 'Be My Baby.' *Sound on Sound*. Available from: <www.soundonsound.com/techniques/classic-tracks-ronettes-be-my-baby> [Accessed 18 November 2020].

Buskin, R. (2009). Donna Summer 'I Feel Love': Classic Tracks. *Sound on Sound*. Available from: <www.soundonsound.com/people/donna-summer-i-feel-love-classic-tracks> [Accessed 18 November 2020].

Butler, M. J. (2014). *Playing with Something That Runs: Technology, Improvisation, and Composition in DJ and Laptop Performance*. New York, Oxford University Press.

Cage, J. (1973). *Silence: Lectures and Writings by John Cage*. Hanover, NH, Wesleyan University Press.

Cage, J. (2015). *Diary: How to Improve the World (You Will Only Make Matters Worse)*. Los Angeles, Siglio Press.

Carr, P. (2018). Once Hollowed, Now Bodied: An Interview with Innovative Electronic Producer Ital Tek. *Pop Matters*. Available from: <www.popmatters.com/ital-tek-bodied-interview-2607285488.html?rebell titem=5#rebelltitem5> [Accessed 18 November 2020].

Carroll, P. (2018). Jon Hopkins Details the Meaning and Method Behind His New Album 'Singularity.' *Stoney Roads*. Available from: <http://stoneyroads.com/2018/04/jon-hopkins-details-the-meaning-and-method-behind-his-new-album-singularity/> [Accessed 18 November 2020].

Catmull, E. (2014). *Creativity, Inc.* New York, Random House.

Cavicchi, D. (1998). *Tramps like Us: Music and Meaning among Springsteen Fans*. New York, Oxford University Press.

Chambliss, D. F. (1989). The Mundanity of Excellence: An Ethnographic Report on Stratification and Olympic Swimmers. *Sociological Theory*, 7 (1), pp. 70–86.

Chanan, M. (1995). *Repeated Takes: A Short History of Recording and its Effects on Music*. London, Verso.

Chanan, M. (1999). *From Handel to Hendrix: The Composer in the Public Sphere*. London, Verso.

Chernoff, J. M. (1979). *African Rhythm and African Sensibility: Aesthetics and Social Action in African Musical Idioms*. Chicago, University of Chicago Press.

Clash Music (2008). Tipper. *Clash Music*. Available from: <www.clashmusic.com/features/tipper> [Accessed 18 November 2020].

Cleveland, B. (2014). Why Modern Audio Recording Might Not Exist without British DIY Audio Pioneer Joe Meek. *TapeOp*. Available from: <https://tapeop.com/articles/100/joe-meek/> [Accessed 18 November 2020].

Clifton, T. (1983). *Music as Heard: A Study in Applied Phenomenology*. New Haven, Yale University Press.

Collins, N., Schedel, M. & Wilson, S. eds. (2013). *Electronic Music*. Cambridge, Cambridge University Press.

Constantinou, S, (2019). Working with Sound in the DAW. In: Hepworth-Sawyer, R., Hodgson, J. & Marrington, M. eds. *Producing Music*. New York, Routledge, pp. 229–245.

Corrigan, C. (2018). Masterclass | Emperor – Sound Design Process. *dBs Music*. Available from: <www.youtube.com/watch?v=9WVVAnDimj8> [Accessed 18 November 2020].

Cosm, T. (2009). An Introduction to Digital Audio Production (Part 1). *Tom Cosm*. Available from: <www.youtube.com/watch?v=Ix2IErEHa7A> [Accessed 18 November 2020].

Costello, S. (2011). Brian Eno's 'Treatments.' *Gearslutz*. Available from: <www.gearslutz.com/board/electronic-music-instruments-and-electronic-music-production/584296-brian-enos-quot-treatments-quot.html> [Accessed 18 November 2020].

Courant, A. (2014). Objekt: Into Another Dimension. *DJMag*. Available from: <https://djmag.com/content/objekt-another-dimension> [Accessed 18 November 2020].

Cummings, P. (1971). Oral History Interview with Robert Motherwell, 1971 Nov. 24–1974 May 1. *Archives of American Art*. Available from: <www.aaa.si.edu/collections/interviews/oral-history-interview-robert-motherwell-13286#transcript> [Accessed 18 November 2020].

Darville, J. (2017). How Actress Made Music Out of Graffiti, Outsider Art, and a Robot. *The Fader*. Available from: <www.thefader.com/2017/04/14/actress-darren-cunningham-azd-interview> [Accessed 18 November 2020].

Day, B. (2010). Mr. Bill – Ableton Tutorial 20: Create a Melody From One MIDI Note. *Mr. Bill*. Available from: <www.youtube.com/watch?v=dgowRYn6z8o> [Accessed 18 November 2020].

Day, B. (2013). Mr. Bill – Speed MIDI Editing in Ableton Live. *Mr. Bill*. Available from: <www.youtube.com/watch?v=2oAX5YxWZug> [Accessed 18 November 2020].

Day, B. (2019). Mr. Bill Masterclass @ BMP College. *Mr. Bill*. Available from: <https://youtu.be/XNbYjPQ9Zkg> [Accessed 18 November 2020].

Day, B. (2020a). Episode 46 – Tom Cosm. *The Mr. Bill Podcast*. Available from: <https://podcasts.apple.com/sn/podcast/tom-cosm/id1480159954?i=1000485546085> [Accessed 18 November 2020].

Day, B. (2020b). Episode 33 – Adam Neely. *The Mr. Bill Podcast*. Available from: <https://podcasts.apple.com/us/podcast/adam-neely/id1480159954?i=1000475724835> [Accessed 18 November 2020].

Day, B. (n.d.). Project Files. *Mr. Bill's Tunes*. Available from: <https://live.mrbillstunes.com/project-files/> [Accessed 18 November 2020].

Deahl, D. (2020). Justin Bieber Was Accused of Stealing a Melody, But It's Actually a Royalty-Free Sample You Can Buy Online. *The Verge*. Available from: <www.theverge.com/2020/2/17/21140838/justin-bieber-changes-running-over-asher-monroe-synergy-splice-sample-melody> [Accessed 18 November 2020].

Demers, J. (2010). *Listening Through the Noise: The Aesthetics of Experimental Electronic Music*. New York, Oxford University Press.

DeSantis, D. (2015). *74 Creative Strategies for Electronic Music Producers*. Berlin, Ableton.

Detrick, B. (2007). The Dirty Heartbeat of the Golden Age. *The Village Voice*. Available from: <www.villagevoice.com/2007/11/06/the-dirty-heartbeat-of-the-golden-age/> [Accessed 23 November 2020].

Didduck, R. A. (2017). *Mad Skills: MIDI and Music Technology in the Twentieth Century*. London, Repeater Books.

Diliberto, J. (1988). Eno Sense. *Music Technology*. Available from: <www.muzines.co.uk/mags/mt/88/02/74> [Accessed 18 November 2020].

Diliberto, J. (2009). Harold Budd & Clive Wright Interview. *Echoes*. Available from: <https://echoes.org/2009/09/11/harold-budd-clive-wright-interview/> [Accessed 18 November 2020].

Diliberto, J. (2016). Harold Budd at 80. *Echoes*. Available from: <https://echoes.org/2016/05/25/harold-budd-at-80/> [Accessed 18 November 2020].

Dodgers, D. (2010). Tipper Interview & Broken Soul Jamboree Preview. *Lost in Sound*. Available from: <http://lostinsound.org/tipper-interview-broken-soul-jamboree-preview/> [Accessed 18 November 2020].

Drake, J. (2018). AI & Music: What Does Artificial Intelligence Mean for Musicians & Producers? *Sound on Sound*. Available from: <www.soundonsound.com/music-business/ai-music> [Accessed 18 November 2020].

Duda, S. (2015). Elite Session with Steve Duda. *Pyramind*. Available from: <www.youtube.com/watch?v=MOUkI5hH2HY&feature=youtu.be> [Accessed 18 November 2020].

Duffet, M. (2013). *Understanding Fandom: An Introduction to the Study of Media Fan Culture*. New York, Bloomsbury Academic.

Duplan, A. (2017). Actress: 'My Saving Grace is I Always Try to Put Beauty in My Music.' *Vice*. Available from: <www.vice.com/en/article/aem34j/actress-new-album-sounds-nothing-like-an-actress-album> [Accessed 18 November 2020].

Durant, A. (1990). A New Day for Music? Digital Technologies in Contemporary Music-Making. In: Hayward, P. ed. *Culture, Technology & Creativity in the Late Twentieth Century*. London, John Libbey, pp. 175–208.

Du Sautoy, M. (2020). *The Creative Code*. Cambridge, MA, Belknap Press.

Eagleman, D. & Brandt, A. (2017). *The Runaway Species*. New York, Catapult.

Earp, J. (2018). William Basinski and the Art of the Infinite. *The Brag*. Available from: <https://thebrag.com/william-basinski-and-the-art-of-the-infinite/> [Accessed 18 November 2020].

Edge, D. (2019). *Cræft: An Inquiry Into the Origins and True Meaning of Traditional Crafts*. New York, W.W. North & Company.

Eede, C. (2017). Youthful Urgency: Lanark Artefax Interviewed. *The Quietus*. Available from: <https://thequietus.com/articles/22622-lanark-artefax-interview> [Accessed 18 November 2020].

Electronic Beats (2016). Jlin's Advice to Aspiring Producers and Her Music-Making Setup. *Electronic Beats*. Available from: <www.electronicbeats.net/native-instruments-komplete-sketches-jlin/ [Accessed 18 November 2020].

Electronic Beats (2019). Actress Explains Why Collaborating with AI Is the Future of Music. *Electronic Beats*. Available from: <www.electronicbeats.net/the-feed/actress-explains-why-collaborating-with-ai-is-the-future-of-music/> [Accessed 18 November 2020].

Eno, B. (1975). *Discreet Music*. London, E.G. Records, Ltd.

Eno, B. (1979). The Studio as Compositional Tool. *Hyperreal*. Available from: <http://music.hyperreal.org/artists/brian_eno/interviews/downbeat79.htm> [Accessed 18 November 2020].

Eno, B. (1996). Generative Music. *inmotionmagazine*. Available from: <https://inmotionmagazine.com/eno1.html> [Accessed 18 November 2020].

Epand, L. (1976). A Phantom Orchestra at Your Fingertips. *Crawdaddy!* Available from: <http://egrefin.free.fr/images/Chamberlin/HCInterview.pdf> [Accessed 18 November 2020].

Eshun, K. (1999). *More Brilliant than the Sun: Adventures in Sonic Fiction*. London, Quartet Books.

FACT Magazine (2016). Rival Consoles – Against the Clock. *FACT Magazine*. Available from: <www. youtube.com/watch?v=B6qe91DX_xk> [Accessed 18 November 2020].

Fallon, P. (2013). In the Studio: Jon Hopkins. *XLR8R*. Available from: <https://xlr8r.com/gear/in-the-studio-jon-hopkins/> [Accessed 18 November 2020].

Fallon, P. (2014). In the Studio: Clark. *XLR8R*. Available from: <https://xlr8r.com/gear/in-the-studio-clark/> [Accessed 18 November 2020].

Fell, M. (2016). Artist Tips: Mark Fell. *XLR8R*. Available from: <https://xlr8r.com/features/artist-tips-mark-fell/> [Accessed 18 November 2020].

Fell, M. (2018). Real Talk: Mark Fell. *XLR8R*. Available from: <https://xlr8r.com/features/real-talk-mark-fell/> [Accessed 18 November 2020].

Felton, D., Scarth, G. & Barker, C. (2018). *The Secrets of Dance Music Production*. Cumbria, Latitude Press.

Finnegan, R. (2007 [1989]). *The Hidden Musicians: Music-Making in an English Town*. Hanover, NH, Wesleyan University Press.

Fintoni, L. (2016). 15 Samplers that Shaped Modern Music. *FACT Magazine*. Available from: <https://www. factmag.com/2016/09/15/15-samplers-that-shaped-modern-music/> [Accessed 18 November 2020].

Firth, S. (1996). *Performing Rites*. Cambridge, MA, Harvard University Press.

Fischer, T. (2012). 15 Questions to Biosphere. *Tokafi*. Available from: <www.tokafi.com/15questions/15-questions-biosphere/> [Accessed 18 November 2020].

Fischer, T. (2014). Interview with Hecq/Ben Lukas Boysen. *Tokafi*. Available from: <www.tokafi. com/15questions/interview-hecq-ben-lukas-boysen/> [Accessed 18 November 2020].

Fischer, T. (n.d.). Fifteen Questions Interview with Guy Sigsworth: Creativity, Fast and Slow. *15 Questions*. Available from: <www.15questions.net/interview/fifteen-questions-interview-guy-sigsworth/page-1/> [Accessed 18 November 2020].

Fisher, M. (2012). Burial: Unedited Transcript. *Wire*. Available from: <www.thewire.co.uk/in-writing/interviews/burial_unedited-transcript> [Accessed 18 November 2020].

Flint, T. (2008). Jean-Michel Jarre 30 Years of Oxygene. *Sound on Sound*. Available from: <www.soundonsound.com/people/jean-michel-jarre> [Accessed 18 November 2020].

Frame, C. (2013). Anatomy of An Enigma: An Interview with Autechre. *The Quietus*. Available from: <https://thequietus.com/articles/13899-autechre-interview-exai-l-event> [Accessed 18 November 2020].

Freeman, P. (2020). Rob Clouth Used the Universe to Make 'Zero Point.' *Bandcamp*. Available from: <https://daily.bandcamp.com/features/rob-clouth-interview> [Accessed 18 November 2020].

Frogs, F. (2014). Ableton Live 9 Tutorial: 30 Quick-Fire Tips. *Point Blank Music School*. Available from: <www.youtube.com/watch?v=KpnOkgsrXP4> [Accessed 18 November 2020].

Frost, T. (2019). Thom Yorke: Daydream Nation. *Crack Magazine*. Available from: <https://crackmagazine. net/article/long-reads/thom-yorke-daydream-nation/> [Accessed 18 November 2020].

Future Music (2011). A Brief History of Steinberg Cubase. *MusicRadar*. Available from: <www.musicradar. com/tuition/tech/a-brief-history-of-steinberg-cubase-406132> [Accessed 18 November 2020].

Future Music (2012). Interview: Skrillex Talks Production, Plug-Ins and Power Edits. *MusicRadar*. Available from: <www.musicradar.com/news/tech/interview-skrillex-talks-production-plug-ins-and-power-edits-545529> [Accessed 18 November 2020].

Future Music (2013). In Pictures: Jon Hopkins' London Studio. *MusicRadar*. Available from: <www. musicradar.com/news/tech/in-pictures-jon-hopkins-london-studio-585726> [Accessed 18 November 2020].

Future Music (2017). How to Create Evolving Ambient Pads with Oscillator Stacking. *MusicRadar*. Available from: <www.musicradar.com/how-to/how-to-create-evolving-ambient-pads-with-oscillator-stacking> [Accessed 4 December 2020].

Future Music (2018a). Jon Hopkins. *Future Music*. Available from: <www.pressreader.com/australia/future-music-9629/20180503/282162176825397> [Accessed 18 November 2020].

Future Music (2018b). Nosaj Thing on Finding His Sound and Working in the Box. *MusicRadar*. Available from: <www.musicradar.com/news/nosaj-thing-on-finding-his-sound-and-working-mostly-in-the-box> [Accessed 18 November 2020].

Future Music (2019). Amon Tobin. *Future Music*. Available from: <www.pressreader.com/australia/future-music-9629/20190502/281505047629215> [Accessed 18 November 2020].

Future Music (2020). Creative Resampling. *MusicRadar*. Available from: <www.magzter.com/article/Music/Future-Music/Creative-Resampling> [Accessed 18 November 2020].

Gaillot, A. D. (2017). For Jlin, the Only Way Is Up. *The Fader*. Available from: <www.thefader.com/2017/05/02/jlin-black-origami-interview> [Accessed 18 November 2020].

Gee, B. (2019). Mr. Bill – His Creative Process. *Chasing Creativity Podcast #44*. Available from: <www.youtube.com/watch?v=sP9LfWwmiQA> [Accessed 18 November 2020].

Gibb, R. (2011). Listening with Your Feet: Boxcutter Interviewed. *The Quietus*. Available from: <https://thequietus.com/articles/06346-boxcutter-interview> [Accessed 18 November 2020].

Gilbert, J. & Pearson, E. (1999). *Discographies: Dance Music, Culture and the Politics of Sound*. London, Routledge.

Girou, B. (2016). Lanark Artefax: Abstract Spaces. *Inverted Audio*. Available from: <https://inverted-audio.com/feature/lanark-artefax-abstract-spaces/> [Accessed 18 November 2020].

Glitch, D. (2019). Arturia Pigments 2! New Granular/Sampler, Buchla-style Filter and Tape Echo. *Das Glitch*. Available from: <www.youtube.com/watch?v=bv5GqHrV5aY> [Accessed 18 November 2020].

Goldmann, S. (2015). *Presets – Digital Shortcuts to Sound*. London, The Bookworm.

Goldstein, D. (1986). The Serpent and the Pearl. *Electronics & Music Maker*. Available from: <www.moredarkthanshark.org/eno_int_eamm-jul86.html> [Accessed 18 November 2020].

Goodwin, A. (1998). Drumming and Memory: Scholarship, Technology, and Music-Making. In: Swiss, T, Sloop, J. & Herman, A. eds. *Mapping the Beat: Popular Music and Contemporary Theory*. Oxford, Blackwell, pp. 121–36.

Gould, G. (1966). The Prospects of Recording. *High Fidelity*. Available from: <https://worldradiohistory.com/Archive-All-Audio/Archive-High-Fidelity/60s/High-Fidelity-1966-04.pdf> [Accessed 18 November 2020].

Grosse, D. (2020). Episode 342 – Stefan Betke (Pole). *Art, Music, Technology Podcast*. Available from: <www.darwingrosse.com/AMT/transcript-0342.html> [Accessed 18 November 2020].

Halle, K. (2010). Interview: Imogen Heap. *Consequence of Sound*. Available from: <https://consequenceofsound.net/2010/08/interview-imogen-heap/> [Accessed 18 November 2020].

Hamer, M. (2005). Interview: Electronic Maestros. *New Scientist*. Available from: <www.newscientist.com/article/mg18524921-400-interview-electronic-maestros/> [Accessed 18 November 2020].

Hamill, J. (2014). The World's Most Famous Electronic Instrument Is Back. Will Anyone Buy the Reissued TB-303? *Forbes*. Available from: <www.forbes.com/sites/jasperhamill/2014/03/25/one-synth-to-rule-them-all-roland-takes-on-clones-with-reissue-of-legendary-tb-303/#1bbd4250359d> [Accessed 18 November 2020].

Hamilton, J. (2016). 808s and Heart Eyes. *Slate*. Available from: <https://slate.com/culture/2016/12/808-the-movie-is-a-must-watch-doc-for-music-nerds.html> [Accessed 18 November 2020].

Hara, M. (2014). Isao Tomita. *Red Bull Music Academy*. Available from: <www.redbullmusicacademy.com/lectures/isao-tomita> [Accessed 18 November 2020].

Harding, P. (2020). *Pop Music Production: Manufactured Pop and BoyBands of the 1990s*. London, Routledge.

Harkins, P. (2020). *Digital Sampling: The Design and Use of Music Technologies*. London, Routledge.

Hauschildt, S. (2019). Fifteen Questions Interview with Steve Hauschildt/Emeralds. *15 Questions*. Available from: <https://15questions.net/interview/fifteen-questions-interview-steve-hauschildt-emeralds/page-1> [Accessed 18 November 2020].

Headphone Commute (2014). In the Studio with Loscil. *Headphone Commute*. Available from: <https://headphonecommute.com/2014/05/26/in-the-studio-with-loscil/> [Accessed 18 November 2020].

Headphone Commute (2016). In the Studio with Ben Lukas Boysen. *Headphone Commute*. Available from: <https://headphonecommute.com/2016/07/27/in-the-studio-with-ben-lukas-boysen/> [Accessed 18 November 2020].

Heaton, A. (2016). Jon Hopkins on DJing, His New Album, and Changing Up Production Styles. *District Magazine*. Available from: <http://districtmagazine.ie/feature/jon-hopkins-on-djing-his-new-album-and-changing-up-production-styles/> [Accessed 18 November 2020].

Hebdige, D. (1987). *"Cut'n' Mix: Culture, Identity and Caribbean Music*. London, Routledge.

Hennion, A. (1989). An Intermediary between Production and Consumption: The Producer of Popular Music. *Technology, & Human Values*, 14 (4), pp. 400–424.

Hertz, T. J. (2012). DSF Q&A 21: Objekt. *Dubstep Forum*. Available from: <www.dubstepforum.com/forum/viewtopic.php?t=251097&start=40> [Accessed 4 December 2020].

Hertz, T. J. (2016). Production w/Objekt v0.1. *Dubstep Forum*. Available from: <www.dubstepforum.com/forum/viewtopic.php   t=251097&sid=8e6515b8bae7f044edd4cd1ddad8504c> [Accessed 18 November 2020].

Hillier, P. (1997). *Arvo Pärt*. Oxford, Oxford University Press.

Hinton, P. (2017). Lanark Artefax's Abstract, Idiosyncratic Sound Design Is Making an Impact. *Mixmag*. Available from: <https://mixmag.net/feature/lanark-artefaxs-abstract-idiosyncratic-sound-design-is-making-an-impact> [Accessed 18 November 2020].

Hislop, J. (2017). Masterclass/Culprate – Creating a Track in Ableton Live. *dBs Music*. Available from: <www.youtube.com/watch?v=BQFStBpRDEM> [Accessed 18 November 2020].

Hodgson, J. (2017). Mix as Auditory Response. In: Hepworth-Sawyer, R. & Hodgson, J. eds. *Mixing Music*. London, Routledge, pp. 216–225.

Hoesen, P. V. (2017). Tech Talk: Peter van Hoesen (Electronic Beats TV). *Telekom Electronic Beats*. Available from: <www.youtube.com/watch?v=GJ4PlMRkIMQ> [Accessed 18 November 2020].

Holkenborg, T. (2020). Finding Your Voice: Sound Design w/Tom Holkenborg (aka Junkie XL). *Junkie XL*. Available from: <www.youtube.com/watch?v=oEO1uLbxnkM> [Accessed 18 November 2020].

Holland, J. H. (2014). *Complexity: A Very Short Introduction*. New York, Oxford University Press.

Holmes, T. (2008). *Electronic and Experimental Music: Technology, Music, and Culture*. London, Routledge.

Huang, A. (2018). 4 Producers Flip the Same Sample FT. Dyalla, Mr. Bill, JVNA. *Andrew Huang*. Available from: <www.youtube.com/watch?v=c8icD9XtRhU> [Accessed 18 November 2020].

Inglis, S. (2003). Four Tet. *Sound on Sound*. Available from: <www.soundonsound.com/people/four-tet> [Accessed 18 November 2020].

Inverted Audio (2016). Lanark Artefax: Abstract Spaces. *Inverted Audio*. Available from: <https://inverted-audio.com/feature/lanark-artefax-abstract-spaces/> [Accessed 18 November 2020].

Izhaki, R. (2017). *Mixing Audio: Concepts, Practices, and Tools*. London, Focal Press.

Jackson, G. (2016). Modern Approaches: Arrangement. *Red Bull Music Academy*. Available from: <https://daily.redbullmusicacademy.com/2016/11/modern-approaches-arrangement> [Accessed 18 November 2020].

Jenkins, H. (2006). *Convergence Culture: Where Old and New Media Collide*. New York, New York University Press.

Jenkinson, T. (2015). Squarepusher Interview – Brutal Sound. *MusicTech*. Available from: <www.musictech.net/features/squarepusher-interview/> [Accessed 18 November 2020].

Kahneman, D. (2013). *Thinking, Fast and Slow*. New York, Farrar, Straus and Giroux.

Keeling, R. (2013). Machine Love: Laurel Halo. *Resident Advisor*. Available from: <www.residentadvisor.net/features/1870> [Accessed 18 November 2020].

Keil, C. (1987). Participatory Discrepancies and the Power of Music. *Cultural Anthropology*, 2 (3), pp. 275–283.

Kelley, F. & Muhammad, A. S. (2015). Hank Shocklee: 'We Had Something to Prove.' *NPR*. Available from: <www.npr.org/sections/microphonecheck/2015/04/16/399817846/hank-shocklee-we-had-something-to-prove> [Accessed 18 November 2020].

Kessler, L. (2014). Interview with Sophie from Elektronauts. *Loukessler*. Available from: <https://loukessler.tumblr.com/post/120118591941/loukessler-interview-with-sophie-from> [Accessed 18 November 2020].

Kirn, P. (2019). Barker, Berghain Resident, Has Found His Voice – and Meaning – in Electronic Sound. *Create Digital Music*. Available from: <https://cdm.link/2019/09/barker-berghain-resident-utility/> [Accessed 18 November 2020].

Kocienda, K. (2018). *Creative Selection*. New York, St. Martin's Press.

Koren, L. (2008). *Wabi-Sabi for Artists, Designers, Poets & Philosophers*. Point Reyes, CA, Imperfect Publishing.

Kraker, L. (2018). Ableton Master G Jones Breaks Down His Creative Approach on the Ineffable Truth [Interview]. *EDM*. Available from: <https://edm.com/interviews/g-jones-interview> [Accessed 18 November 2020].

Kretowicz, S. (2016). 'I Am Lost with Infinite Choices': Tim Hecker on the Information Overload of His New Album Love Streams. *FACT Magazine*. Available from: <www.factmag.com/2016/03/31/tim-hecker-interview-love-streams/> [Accessed 18 November 2020].

Kubik, G. (1962). The Phenomenon of Inherent Rhythms in East and Central African Instrumental Music. *African Music*, 3 (1), pp. 33–42.

Lagomarsino, J. (2017). The Skeuomorphic Hell of Music-Making Apps. *BoingBoing*. Available from: <https://boingboing.net/2017/08/25/the-skeuomorphic-hell-of-music.html> [Accessed 18 November 2020].

Lamar, C. (2019). 'Old Town Road' Producer YoungKio on How Lil Nas X's Song Came to Life. *Billboard*. Available from: <www.billboard.com/articles/columns/hip-hop/8504409/old-town-road-producer-youngkio-interview-lil-nas-x> [Accessed 18 November 2020].

Lanier, J. (2010). *You Are Not a Gadget*. New York, Vintage.

Lavengood, M. (2019). What Makes It Sound '80s? The Yamaha DX7 Electric Piano Sound. *Journal of Popular Music Studies*, 31 (3), pp. 73–94.

Laxity (2020). To Put a Lid on This Whole Issue. *Twitter*. Available from: <https://twitter.com/laxcitymusic/status/1228487508314423296?lang=en> [Accessed 4 December 2020].

Lea, T. (2014). Watching Westerns with the Sound Off: Actress Interviewed. *FACT Magazine*. Available from: <www.factmag.com/2014/02/07/actress-interview-fact-ghettoville-darren-cunningham/> [Accessed 18 November 2020].

Lee, I. (1997). Ted Macero Interview. *Perfect Sound Forever*. Available from: <https://www.furious.com/perfect/teomacero.html> [Accessed 18 November 2020].

Leski, K. (2015). *The Storm of Creativity*. Cambridge, MA, MIT Press.

Lidell, J. (2020). Episode 72 – Olivier Alary and Johannes Malfatti. *Hanging Out with Audiophiles Podcast*. Available from: <http://jamielidellmusic.com/new-page-2> [Accessed 18 November 2020].

Lopatin, D. (2017). Electric Independence: Daniel Lopatin (Oneohtrix Point Never) – Motherboard. *YouTube*. Available from: <www.youtube.com/watch?v=E0RAmNU5Es8> [Accessed 18 November 2020].

Lord, N. (1989). Steinberg Cubase Atari ST Software, Part 1. *Music Technology*. Available from: <www.muzines.co.uk/articles/steinberg-cubase/121> [Accessed 18 November 2020].

Lubner, R. (2019). Logos. *Tiny Mix Tapes*. Available from: <www.tinymixtapes.com/features/logos> [Accessed 18 November 2020].

Lynch, D. (2007). *Catching the Big Fish: Meditation, Consciousness, and Creativity*. New York, TarcherPerigee.

M., J. (2019). An Interview with Loscil. *Boston Hassle*. Available from: <https://bostonhassle.com/an-interview-with-loscil/> [Accessed 18 November 2020].

Magnusson, T. (2019). *Sonic Writing: Technologies of Material, Symbolic & Signal Inscriptions*. New York, Bloomsbury.

Mao, J. (2013a). Q-Tip. *Red Bull Music Academy*. Available from: <www.redbullmusicacademy.com/lectures/q-tip> [Accessed 18 November 2020].

Mao, J. (2013b). Questlove. *Red Bull Music Academy*. Available from: <www.redbullmusicacademy.com/lectures/questlove-new-york-2013> [Accessed 18 November 2020].

Marrington, M. (2011). Experiencing Musical Composition in the DAW: The Software Interface as Mediator of the Musical Idea. *Journal on the Art of Record Production*. Available from: <www.arpjournal.com/asarpwp/experiencing-musical-composition-in-the-daw-the-software-interface-as-mediator-of-the-musical-idea-2/> [Accessed 18 November 2020].

Marrington, M. (2019). The DAW, Electronic Music Aesthetics, and Genre Transgression in Music Production: The Case of Heavy Metal Music. In: Hepworth-Sawyer, R., Hodgson, J. & Marrington, M. eds. *Producing Music*. London, Routledge, pp. 52–74.

Marsden, R. (2008). Rhythm King: The Return of the Roland 808 Drum Machine. *Independent*. Available from: <www.independent.co.uk/arts-entertainment/music/ features/rhythm-king-the-return-of-the-roland-808-drum-machine-1066808.html> [Accessed 18 November 2020].

Martin, L. (2018). Objekt. *Red Bull Music Academy*. Available from: <www.redbullmusicacademy.com/lectures/objekt> [Accessed 18 November 2020].

Matthews, M. V. (1963). The Digital Computer as a Musical Instrument. *Science, New Series*, 142 (3592), pp. 553–557.

McDermott, M. (2020). EX. 526: John Frusciante. *RA Exchange Podcast*. Available from: <www.residentadvisor.net/podcast-episode.aspx?exchange=526> [Accessed 18 November 2020].

McNeill, M. (2015). Interview: Daniel Lanois. *Red Bull Music Academy*. Available from: <https://daily.redbullmusicacademy.com/2015/10/daniel-lanois-interview> [Accessed 18 November 2020].

Meadows, D. H. (2008). In: Wright, D. ed. *Thinking in Systems: A Primer*. New York, Chelsea Green Publishing.

Merrich, J. (n.d.). Caterina Barbieri: New Tactics for Electronic Mutants. *Digicult*. Available from: <http://digicult.it/news/caterina-barbieri-new-tactics-for-electronic-mutants/> [Accessed 18 November 2020].

Metz, C. (2020). Meet GPT-3. It Has Learned to Code (and Blog and Argue). *The New York Times*. Available from: <www.nytimes.com/2020/11/24/science/artificial-intelligence-ai-gpt3.html> [Accessed 25 November 2020].

Meyer-Horn, M. (2019). Interview: Ólafur Arnalds Is the Enfant Terrible of Modern Music. *Enfntsterribles*. Available from: <https://enfntsterribles.com/interview-olafur-arnalds-is-the-enfant-terrible-of-modern-music/> [Accessed 18 November 2020].

Miller, K. (2012). *Playing Along: Digital Games, YouTube, and Virtual Performance*. New York, Oxford University Press.

Mills, C. W. (2000). On Intellectual Craftsmanship. In: *The Sociological Imagination*. New York, Oxford University Press, pp. 195–226.

Milner, G. (2010). *Perfecting Sound Forever: An Aural History of Recorded Music*. New York, Farrer, Straus and Giroux.

Minsker, E. (2014). Aphex Twin's List of Gear Used on *Syro* Surfaces. *Pitchfork*. Available from: <https://pitchfork.com/news/56630-aphex-twins-list-of-gear-used-on-syro-surfaces/> [Accessed 18 November 2020].

Modartt (2020). Pianoteq 7 – Modartt introduces Acoustic Morphing. *YouTube*. Available from: <www.youtube.com/watch?v=k2oZlMueJCM&feature=youtu.be> [Accessed 18 November 2020].

Montero, B. G. (2016). *Thought in Action: Expertise and the Conscious Mind*. New York, Oxford University Press.

Moody, R. (2012). Europe: Forsake Your Drum Machines! A Genealogy. In: *On Celestial Music: And Other Adventures in Listening*. New York, Back Bay Books, pp. 352–417.

Moorefield, V. (2005). *The Producer as Composer*. Cambridge, MA, MIT Press.

Morley, P. (2005). *Words and Music: A History of Pop in the Shape of a City*. London, Bloomsbury.

Morse, E. (2015). Arca Talks Working with Bjork, Screaming About Sex, Explosive New LP. *Rolling Stone*. Available from: <www.rollingstone.com/music/music-news/arca-talks-working-with-bjork-screaming-about-sex-explosive-new-lp-41272/> [Accessed 18 November 2020].

Moylan, W. (2017). How to Listen, What to Hear. In: Hepworth-Sawyer, R., Hodgson, J. & Marrington, M. eds. *Mixing Music*. London, Routledge, pp. 24–52.

Moylan, W. (2020). *Recording Analysis: How the Record Shapes the Song*. London, Routledge.

Muggs, J. (2016). Autechre: Elseq et al. *Resident Advisor*. Available from: <www.residentadvisor.net/features/2756> [Accessed 18 November 2020].

Müller, B. (2017). Feature: Skee Mask (Electronic Beats TV). *Telekom Electronic Beats*. Available from: <www.youtube.com/watch?v=jKtJKeBRKuY> [Accessed 18 November 2020].

Multiplier (2017). Au5 in the DAW | Arise | Dubstep in Ableton Live. *Multiplier*. Available from: <www.youtube.com/watch?v=XiQr-lO2obs> [Accessed 18 November 2020].

Mumford, M. D., Medeiros, K. E. & Partlow, P. J. (2012). Creative Thinking: Processes, Strategies, and Knowledge. *The Journal of Creative Behavior*, 46 (1), pp. 30–47.

Murphy, J. (2016). Noisia Reveal Some of Their Best Production Techniques in New Interview with Point Blank. *Raverrafting*. Available from: <http://raverrafting.com/point-blank-nosia-interview/2016/07/28/> [Accessed 18 November 2020].

MusicRadar (2019). Jon Hopkins: 'What Do I Call My Music? Beats with Melodies.' *MusicRadar*. Available from: <https://www.musicradar.com/news/jon-hopkins-what-do-i-call-my-music-beats-with-melodies> [Accessed 18 November 2020].

Myson, A. (2011). Ital Tek (Planet Mu/Atom River) @ Dubspot – Interview, Workshop Recap + New EP 'Gonga'. *YouTube*. Available from: <www.youtube.com/watch?v=uCj6J6TRqqQ> [Accessed 18 November 2020].

Nagshineh, A. (2013). Cover Story: Flying Lotus Interview. *Bonafide*. Available from: <www.bonafidemag.com/cover-story-flying-lotus-interview/> [Accessed 18 November 2020].

ncross10 (2020). Virtual Riot Making a Song from Start to Finish! [Reupload]. *ncross10*. Available from: <www.youtube.com/watch?v=ed-2NLGQA8U&t=2s> [Accessed 18 November 2020].

Neill, B. (2004). Breakthrough Beats: Rhythm and the Aesthetics of Contemporary Electronic Music. In: Cox, C. & Warner, D. eds. *Audio Culture: Readings in Modern Music*. London, Continuum, pp. 386–391.

Nguyen, D. V. (2018). Jon Hopkins: Singularity Review – Beautiful but Heartbreaking. *Irish Times*. Available from: <www.irishtimes.com/culture/music/jon-hopkins-singularity-review-beautiful-but-heartbreaking-1.3479292> [Accessed 18 November 2020].

Nicolai, C. (2018). Ask the Experts: Alva Noto. *XLR8R*. Available from: <https://xlr8r.com/features/ask-the-experts-alva-noto/> [Accessed 18 November 2020].

Nowness (2011). Gerhard Richter Painting: Watch the Master Artist at Work. *Nowness*. Available from: <www.youtube.com/watch?v=yF6EluMNR14> [Accessed 18 November 2020].

O'Gara, C. (2019). Interview – Stenny – 'Not Breakbeat, Although The Beats Are Broken.' *Monument*. Available from: <https://mnmt.no/magazine/2019/11/29/interview-stenny-not-breakbeat-although-the-beats-are-broken/> [Accessed 18 November 2020].

Oram, D. (2016). *An Individual Note of Music, Sound and Electronics*. London, Anomie Publishing.

Oskillator (2017a). Deleted Deadmau5 Chord Progressions. *Oskillator*. Available from: <www.youtube.com/watch?v=p5rXzLubwGA> [Accessed 18 November 2020].

Oskillator (2017b). Deadmau5 Makes THAT SYNTH SOUND . . . and Junks it. *Oskillator*. Available from: <www.youtube.com/watch?v=V2FT34ZBN2E> [Accessed 18 November 2020].

Pareles, J. (2013). Brothers Who Make Electronica by Hand. *The New York Times*. Available from: <www.nytimes.com/2013/06/12/arts/music/tomorrows-harvest-by-boards-of-canada.html> [Accessed 18 November 2020].

Pareles, J. (2020). Autechre Worked in Isolation for Decades. Now It's Unintentionally Timely. *The New York Times*. Available from: <www.nytimes.com/2020/10/13/arts/music/autechre-sign-interview.html?referringSource=articleShare> [Accessed 18 November 2020].

Pattison, L. (2013). Boards of Canada: 'We've Become a Lot More Nihilistic Over the Years.' *The Guardian*. Available from: <www.theguardian.com/music/2013/jun/06/boards-of-canada-become-more-nihilistic> [Accessed 18 November 2020].

Payne, O. (2019). London Producer O'Flynn on Bringing Humanity to Electronic Music. *MusicTech*. Available from: <www.musictech.net/features/interviews/oflynn-aletheia-interview/> [Accessed 18 November 2020].

Pazienti-Caiden, M. (2020). EX 523: Ikonika. *RA Podcast*. Available from: <www.residentadvisor.net/podcast-episode.aspx?exchange=524> [Accessed 18 November 2020].

Peel, I. (2005). Trevor Horn: 25 Years of Hits. *Sound on Sound*. Available from: <www.soundonsound.com/people/trevor-horn> [Accessed 18 November 2020].

Pequeno, Z. (2010). Autechre. *tinymixtapes*. Available from: <https://www.tinymixtapes.com/features/autechre> [Accessed 18 November 2020].

Petit, P. (2015). *Creativity: The Perfect Crime*. New York, Riverhead Books.

Petruska, N. (2017). Sound Design 04: Serum. *Frequent*. Available from: <www.youtube.com/watch?v=FG8e4EYXJsM&t=1183s> [Accessed 18 November 2020].

Petruska, N. (2020). Arrangement 03: Reconstructing and Expanding. *Frequent*. Available from: <www.youtube.com/watch?v=EsRnJ8bCX-o> [Accessed 18 November 2020].

Pinch, T. & Trocco, F. (2004). *Analog Days: The Invention and Impact of the Moog Synthesizer*. Cambridge, MA, Harvard University Press.

Polyani, M. (1966). *The Tacit Dimension*. New York, Doubleday & Company.

Prior, N. (2008). Ok Computer: Mobility, Software and the Laptop Musician. *Information, Communication & Society*, 11 (7), pp. 912–932.

Prior, N. (2009). Software Sequencers and Cyborg Singers: Popular Music in the Digital Hypermodern. *New Formations*, 66, pp. 81–99.

Prior, N. (2018). *Popular Music: Digital Technology and Society*. London, Sage.

Punathambekar, A. (2007). Between Rowdies and Rasikas: Rethinking Fan Activity in Indian Film Culture. In: Gray, J., Sandvoss, C. & Harrington, C. L. eds. *Fandom: Identities and Communities in a Mediated World*. New York, New York University Press, pp. 198–209.

Pyramind (2020). Mr. Bill's Guide to Sound Design Mud Pies. *Pyramind*. Available from: <https://pyramind.com/mr-bills-guide-to-sound-design-mud-pies/> [Accessed 18 November 2020].

Raihani, N. (2017). How to Get Through the Frustrations of Making Music. *Electronic Beats*. Available from: <www.electronicbeats.net/native-instruments-komplete-sketches-deru/> [Accessed 18 November 2020].

Ralston, W. (2019). Barker: Moving with Melody. *XLR8R*. Available from: <https://xlr8r.com/features/barker-moving-with-melody/> [Accessed 18 November 2020].

Ramage, M. (2010). History of Cubase. *Exxos*. Available from: <www.exxoshost.co.uk/forum/viewtopic.php?t=319> [Accessed 18 November 2020].

Rancic, M. (2018). Jon Hopkins on the Tension in Blending Techno and Ambient Sounds for His Latest Album 'Singularity.' *Uproxx*. Available from: <https://uproxx.com/music/jon-hopkins-singularity-interview/> [Accessed 18 November 2020].

Ransby (2020). How to Create Your Own Instruments. *Ableton*. Available from: <https://youtu.be/fQdo6HsBbJ0> [Accessed 18 November 2020].

Ravens, C. (2019). Beatrice Dillon: The Most Thrilling New Artist in Electronic Music. *The Guardian*. Available from: <www.theguardian.com/music/2019/dec/27/beatrice-dillon-the-most-thrilling-new-artist-in-electronic-music> [Accessed 18 November 2020].

Reddit (2017). Tipper Production Methods/Ethics Thread. *Reddit*. Available from: <www.reddit.com/r/Tipper comments/5nhgon/tipper_production_methodsethics_thread/> [Accessed 18 November 2020].

Reddit (2018). Who Are the Most Technically Talented Producers in EDM/Electronic Music? *Reddit*. Available from: <www.reddit.com/r/electronicmusic/comments/9isja8/who_are_the_most_technically_talented_producers/> [Accessed 18 November 2020].

Reddit (2019). What Are Some Tips to Make the Arrangement Stand Out? *Reddit*. Available from: <www.reddit.com/r/edmproduction/comments/9zrjgd/discussion_what_are_some_tips_to_make_the/> [Accessed 18 November 2020].

Repeatle (2010). Akai S-950 Resampling (Enough Already!). *Repeatle*. Available from: <www.youtube.com/watch?v=-YmOj1vL-zY> [Accessed 18 November 2020].

RetroSynthAds (2019). Steinberg Cubit 'Visual Song Processing' Advertisement, Keyboard 1989 / Cubase 'Buy it! Boot it! Love it!' Advertisement, Electronic Musician 1990. *Retrosynthads*. Available from: <http://retrosynthads.blogspot.com/2019/05/steinberg-cubit-visual-song-processing.html> [Accessed 18 November 2020].

Reynolds, S. (1999). *Generation Ecstasy: Into the World of Techno and Rave Culture*. London, Routledge.

Reynolds, S. (2017). Why Burial's Untrue Is the Most Important Electronic Album of the Century So Far. *Pitchfork*. Available from: <https://pitchfork.com/features/article/why-burials-untrue-is-the-most-important-electronic-album-of-the-century-so-far/> [Accessed 18 November 2020].

Roads, C. (2015). *Composing Electronic Music: A New Aesthetic*. New York, Oxford University Press.

Robinson, P. (2019). Masterclass: Philth on Finding Inspiration and Developing Your Workflow. *dBs Music*. Available from: <www.youtube.com/watch v=F8CVUHRaYrg&feature=youtu.be> [Accessed 18 November 2020].

Rubik, E. (2020). *Cubed: The Puzzle of Us All*. New York, Flatiron Books.

Ryce, A. (2016). Biosphere: Introverted Music. *Resident Advisor*. Available from: <www.residentadvisor.net/features/2805> [Accessed 18 November 2020].

Sadowick, B. (2014). Basic Track Arrangement & Tips in Ableton Live. *SadowickProduction*. Available from: <www.youtube.com/watch?v=hH-Efw9m2iY> [Accessed 18 November 2020].

Salvatier, J. (2017). Reality Has a Surprising Amount of Detail. *John Salvatier*. Available from: <http://johnsalvatier.org/blog/2017/reality-has-a-surprising-amount-of-detail> [Accessed 18 November 2020].

Samplesbank (2008). How do J Dilla, Madlib, and a Few Others Get That off Beat Sound? *Future Producers*. Available from: <www.futureproducers.com/forums/production-techniques/getting-started/how-do-j-dilla-madlib-few-others-get-off-beat-sound-240999/> [Accessed 18 November 2020].

Sanders, T. (2017). J Dilla & the Shape of Jazz to Come. *Medium*. Available from: <https://medium.com/@ tomcsanders/j-dilla-the-shape-of-jazz-to-come-164b2697d97> [Accessed 18 November 2020].

Saydlowski, B. (1982). The A.F.M. *Modern Drummer*, February/March, p. 20.

Schaeffer, P. (2012). *In Search of a Concrete Music*. North, C. & Dack, J., trans. Berkeley, University of California Press.

Scheps, A. (2019). Mixing on Headphones. *Gearslutz*. Available from: <www.gearslutz.com/board/q-a-with-andrew-scheps/1261637-mixing-headphones.html> [Accessed 18 November 2020].

Schmidt, T. (2003). Prince Paul. *Red Bull Music Academy*. Available from: <www.redbullmusicacademy. com/lectures/prince-paul-prince-of-thieves> [Accessed 18 November 2020].

Schmidt, T. (2014). Dave Smith. *Red Bull Music Academy*. Available from: <www.redbullmusicacademy. com/lectures/dave-smith-tokyo-2014> [Accessed 18 November 2020].

Seabrook, J. (2016). How Mike Will Made It. *The New Yorker*. Available from: <www.newyorker.com/ magazine/2016/07/11/how-mike-will-made-it> [Accessed 18 November 2020].

Shapiro, P. (2015). *Turn the Beat Around: The Secret History of Disco*. New York, Farrar, Straus and Giroux.

Shelvock, M. T. (2020). *Cloud-Based Music Production*. London, Routledge.

Sherburne, P. (2014). Strange Visitor: A Conversation with Aphex Twin. *Pitchfork*. Available from: <https:// pitchfork.com/features/cover-story/reader/aphex-twin/> [Accessed 18 November 2020].

Sherburne, P. (2018). Autechre on Their Epic NTS Sessions, David Lynch, and Where Code Meets Music. *Pitchfork*. Available from: <https://pitchfork.com/thepitch/autechre-interview-nts-sessions-david-lynch-where-code-meets-music/> [Accessed 18 November 2020].

Simpson, D. (2018a). How We Made: Roni Size on the Mercury-winning Album New Forms. *The Guardian*. Available from: <www.theguardian.com/music/2018/jul/17/how-we-made-roni-size-on-the-mercury-winning-album-new-forms> [Accessed 18 November 2020].

Simpson, D. (2018b). More Synthetic Bamboo! The Greatest Preset Sounds in Pop Music. *The Guardian*. Available from: <www.theguardian.com/music/2018/aug/14/the-greatest-preset-sounds-in-pop-music> [Accessed 18 November 2020].

Smith, K. A. (2016). Kaitlyn Aurelia Smith – in the Studio. *FACT Magazine*. Available from: <www.you tube.com/watch?v=95UvPlhjbE4> [Accessed 18 November 2020].

Smith, L. (2018a). Objekt. *Dazed Digital*. Available from: <www.dazeddigital.com/music/article/13367/1/ objekt> [Accessed 18 November 2020].

Smith, M. (2018b). Machine Love: Jon Hopkins. *Resident Advisor*. Available from: <www.residentadvisor. net/features/3210> [Accessed 18 November 2020].

Smith, M. (2019). The Art of Production: Floating Points. *Resident Advisor*. Available from: <www.residen-tadvisor.net/features/3548> [Accessed 18 November 2020].

Snead, J. A. (1984). Repetition as a Figure of Black Culture. In: Gates, H. L. Jr. ed. *Black Literature and Literary Theory*. New York, Methuen, pp. 59–79.

Snider, G. (2017). *The Shape of Ideas: An Illustrated Exploration of Creativity*. New York, Harry N. Abrams.

Spitfire Audio (n.d.a). Ólafur Arnalds Evolutions. *Spitfire Audio*. Available from: <www.spitfireaudio.com/ shop/a-z/olafur-arnalds-chamber-evolutions/> [Accessed 18 November 2020].

Spitfire Audio (n.d.b). Ólafur Arnalds Composer Toolkit. *Spitfire Audio*. Available from: <Spitfire Audio (n.d.). Ólafur Arnald> [Accessed 18 November 2020].

Splice Blog (2020). Exclusive: Laxity on the Moment he Heard His Sample on Justin Bieber's 'Running Over.' *Splice*. Available from: <https://splice.com/blog/laxity-justin-bieber-running-over/> [Accessed 18 November 2020].

Stancazk, J. (2020). Episode 47 – Kill The Noise. *The Mr. Bill Podcast*. Available from: <https://podcasts. apple.com/us/podcast/kill-the-noise/id1480159954?i=1000486293497> [Accessed 18 November 2020].

Stationary Travels (2018). Duologue: A Conversation with r Beny. *Stationary Travels*. Available from: <https://stationarytravels.wordpress.com/2018/03/04/duologue-a-conversation-with-r-beny/> [Accessed 18 November 2020].

Strange, A. (1971). *Electronic Music: Systems, Techniques, and Controls*. Dubuque, IA, William C Brown.

Stratchan, R. (2017). *Sonic Technologies: Popular Music, Digital Culture and the Creative Process*. New York, Bloomsbury Academic.

Sudnow, D. (2001). *Ways of the Hand*. Cambridge, MA, MIT Press.

swc (2017). An Interview with Steve Hauschildt at the Green Elephant in Dallas, TX. *Somewherecold*. Available from: <https://somewherecold.net/2017/07/03/an-interview-with-steve-hauschildt-at-the-green-elephant-in-dallas-tx-june-27–2017/> [Accessed 18 November 2020].

Taleb, N. N. (2014). *Antifragile: Things That Gain from Disorder*. New York, Random House.

Théberge, P. (1997). *Any Sound You Can Imagine*. Hanover, NH, Wesleyan University Press.

Tingen, P. (2004). Autechre: Recording Electronica. *Sound on Sound*. Available from: <www.soundonsound.com/people/autechre> [Accessed 18 November 2020].

Tingen, P. (2017). Producing EDM: The Zedd Interview. *Sound on Sound*. Available from: <www.soundonsound.com/people/producing-edm> [Accessed 18 November 2020].

Tjora, A. H. (2009). The Groove in the Box: A Technologically Mediated Inspiration in Electronic Dance Music. *Popular Music*, 28 (2), pp. 161–177.

Toland, J. (2009). Interview: Pole. *FACT Magazine*. Available from: <www.factmag.com/2009/01/01/interview-pole/> [Accessed 18 November 2020].

Tomkins, C. (2019). Vija Celmins's Surface Matters. *The New Yorker*. Available from: <www.newyorker.com/magazine/2019/09/02/vija-celmins-surface-matters> [Accessed 18 November 2020].

Toop, D. (1999). *Rap Attack 3: African Rap to Global Hip Hop*. London, Serpent's Tail.

Toop, D. (2011). A-Z of Electro. *Wire*. Available from: <www.thewire.co.uk/in-writing/essays/a-z-of-electro> [Accessed 18 November 2020].

Tsugi (2018). Autechre's Tsugi Interview Translated. *Reddit*. Available from: <www.reddit.com/r/autechre/comments/9of3dv/autechres_tsugi_interview_translated/> [Accessed 18 November 2020].

Turner, D. (2020a). Ital Tek. *FutureMusic*. Available from: <www.magzter.com/article/Music/Future-Music/Ital-Tek> [Accessed 18 November 2020].

Turner, D. (2020b). Beatrice Dillon. *FutureMusic*. Available from: <www.magzter.com/article/Music/Future-Music/Beatrice-Dillon> [Accessed 18 November 2020].

Twells, J. (2015). Lorenzo Senni Talks Hardcore, Cyberpunk and Appropriating Dance Music's Most Reviled Genre. *FACT Magazine*. Available from: <www.factmag.com/2015/02/23/trance-im-trance-now-lorenzo-senni-talks-hardcore-cyberpunk-appropriating-dance-musics-reviled-genre/> [Accessed 18 November 2020].

Twells, J. (2016). The 14 Pieces of Software that Shaped Modern Music. *FACT Magazine*. Available from: <https://www.factmag.com/2016/10/01/the-14-pieces-of-software-that-shaped-modern-music/> [Accessed 18 November 2020].

Villalobos, R. & Loderbauer, M. (2014). Ricardo Villalobos/Max Loderbauer Talk about 'Re: ECM' (Interview). *ECM Records*. Available from: <www.youtube.com/watch?v=hxMWIrbWmsM> [Accessed 18 November 2020].

Voorn, J. (2016). Tech Talk: Joris Voorn (Electronic Beats TV). *Telekom Electronic Beats*. Available from: <www.youtube.com/watch?v=N2Vqt4bEnJo> [Accessed 18 November 2020].

Waksman, S. (2004). California Noise: Tinkering with Hardcore and Heavy Metal in Southern California. *Social Studies of Science*, 24 (5), pp. 675–702.

Walden, J. (2010). Project Templates: Steinberg Cubase Tips & Techniques. *Sound on Sound*. Available from: <www.soundonsound.com/techniques/project-templates> [Accessed 18 November 2020].

Waldorf. (2021). "4-Pole." *Waldorf Music*. Available from: <https://waldorfmusic.com/en/hardware-archive/4-pole> [Accessed 18 November 2020].

Walmsley, D. (2009). Dubstep. In: Young, R. ed. *The WIRE Primers: A Guide to Modern Music*. London, Verso.

Walmsley, D. (2010). Monolake in Full. *Wire*. Available from: <www.thewire.co.uk/in-writing/interviews/monolake-in-full> [Accessed 18 November 2020].

Warner, D. (2017). *Live Wires: A History of Electronic Music*. London, Reaktion Books.

Warner, T. (2003). *Pop Music – Technology and Creativity: Trevor Horn and the Digital Revolution*. London, Ashgate.

Weschler, L. (2012). Uncanny Valley: On the Digital Animation of the Face. In: *Uncanny Valley and Other Adventures in the Narrative*. Berkeley, CA, Counterpoint, pp. 1–20.

West, R. L. (2018). Rival Consoles, Reddit AMA. *Reddit*. Available from: <www.reddit.com/r/electronicmusic/comments/9366hc/rival_consoles_ama/> [Accessed 18 November 2020].

West, R. L. (2020a). Final Arrangement of 'Hidden' from Persona LP. *Twitter*. Available from: <https://twitter.com/rivalconsoles/status/1252370499998384130> [Accessed 18 November 2020].

West, R. L. (2020b). Diagram. *Rival Consoles*. Available from: <www.rivalconsoles.net/diagram> [Accessed 18 November 2020].

White, P. (2006). Mixing Essentials. *Sound on Sound*. Available from: <www.soundonsound.com/techniques/mixing-essentials> [Accessed 18 November 2020].

Wilson, S. (2016). The 14 Drum Machines That Shaped Modern Music. *FACT Magazine*. Available from: <www.factmag.com/2016/09/22/the-14-drum-machines-that-shaped-modern-music/> [Accessed 18 November 2020].

Wilson, S. (2017). Four Tet's Studio Just Inspired the Month's Best Twitter Meme. *FACT Magazine*. Available from: <https://www.factmag.com/2017/10/27/four-tet-studio-meme/> [Accessed 18 November 2020].

Wilson, S. (2018a). Signal Path: Caterina Barbieri on Synthesis, Minimalism and Creating Living Organisms Out of Sound. *FACT Magazine*. Available from: <www.factmag.com/2018/07/08/caterina-barbieri-signal-path/> [Accessed 18 November 2020].

Wilson, S. (2018b). Signal Path: Mark Fell on His Love of FM Synthesis and Algorithmic Composition. *FACT Magazine*. Available from: <www.factmag.com/2018/10/06/mark-fell-signal-path/> [Accessed 18 November 2020].

Wilson, S. (2019). Logos on the Classic Methods Behind His Genre-Bending Club and Ambient Experiments. *FACT Magazine*. Available from: <www.factmag.com/2019/06/14/logos-signal-path/> [Accessed 18 November 2020].

Worrall, D. (2019). Beginner's Guide to Saturation. *fabfilter*. Available from: <www.youtube.com/watch?v=NO2OZ3UTy2k> [Accessed 18 November 2020].

Yoo, N. (2017). Holly Herndon and Jlin Discuss Production and Creative Process in New Interview. *Pitchfork*. Available from: <https://pitchfork.com/news/holly-herndon-and-jlin-discuss-production-and-creative-process-in-new-interview/> [Accessed 18 November 2020].

Young, S. (2003). Amon Tobin: Out from Out Where. *Sound on Sound*. Available from: <www.soundonsound.com/people/amon-tobin> [Accessed 18 November 2020].

Zagorski-Thomas, S. (2014). *The Musicology of Record Production*. Cambridge, Cambridge University Press.

Ziemelis, K. & Allen, L. (2001). Complex Systems. *Nature*. Available from: <www.nature.com/articles/35065672> [Accessed 18 November 2020].

Zimmerman, J. (2015). Deadmau5. *Twitch*. Available from: <www.twitch.tv/videos/62852951> [Accessed 18 November 2020].

Zynaptic (n.d.). Morph 2.0. *Zynaptic*. Available from: <www.zynaptiq.com/morph/> [Accessed 26 November 2020].

Index

Printed in the United States
by Baker & Taylor Publisher Services